Electrical Systems 1

Electrical Systems 1

From Diagnosis to Prognosis

Edited by

Abdenour Soualhi
Hubert Razik

iSTE

WILEY

First published 2020 in Great Britain and the United States by ISTE Ltd and John Wiley & Sons, Inc.

ISTE Ltd
27-37 St George's Road
London SW19 4EU
UK

www.iste.co.uk

John Wiley & Sons, Inc.
111 River Street
Hoboken, NJ 07030
USA

www.wiley.com

Library of Congress Control Number: 2019955247

British Library Cataloguing-in-Publication Data
A CIP record for this book is available from the British Library
ISBN 978-1-78630-465-0

MIX
Paper from responsible sources
FSC
www.fsc.org
FSC® C013604

Contents

Introduction

The diagnosis and prognosis of electrical systems is still a relevant field of research. The research that has been carried out over the years has made it possible to acquire enough knowledge, to build a base from which we can delve further into this field of research. This study is a new challenge that estimates the remaining lifetime of the analyzed process. Many studies have been carried out to establish a diagnosis of the state of health of an electric motor, for example. However, making a diagnosis is like giving binary information: the condition is either healthy or defective. Of course, this may seem simplistic, but detecting a failure requires the use of suitable sensors that provide signals. These will be processed to monitor health indicators (features) for defects. Then, we witnessed a multitude of research activities around classification. It was indeed appropriate to distinguish the operating states, to differentiate them from one another and to inform the operator of the level of severity of a failure or even of the type of failure among a predefined panel. A major effort has been made to estimate the remaining lifetime or even the lifetime consumed. This is a challenge that many researchers are still trying to meet.

This book, which has been divided into two volumes, informs readers about the theoretical approaches and results obtained in different laboratories in France and also in other countries such as Spain, and so on. To this end, many researchers from the scientific community have contributed to this book by sharing their research results.

Introduction written by Abdenour SOUALHI and Hubert RAZIK.

Chapter 1, Volume 1, "Diagnostic Methods for the Health Monitoring of Gearboxes", by A. Soualhi and H. Razik, presents state-of-the-art diagnostic methods used to analyze the defects present in gearboxes. First of all, there is a bibliographical presentation regarding different types of gears and their defects. We conclude that gear defects represent the predominant defect at this level, thus justifying the interest in detecting and diagnosing them. Then, we present various gear analyses and monitoring techniques proposed as part of the condition-based maintenance and propose a diagnostic method. Thus, we show the three main phases of diagnosis: First, the analysis presented as a set of technical processes ensuring control of the representative quantities of operation; then the monitoring that exploits the fault indicators for detection; finally, the diagnosis which is the identification of the detected defect.

Chapter 2, Volume 1, "Techniques for Predicting Defects in Bearings and Gears", by A. Soualhi and H. Razik, deals with strategies based on features characterizing the health status of the system to predict the appearance of possible failures. The prognosis of faults in a system means the prediction of the failure imminence and/or the estimation of its remaining life. It is in this context that we propose, in this chapter, the three methods of prognosis. In the first method, the degradation process of each system is modeled by a hidden Markov model (HMM). In a measured sequence of observations, the solution consists of identifying among the HMMs the one that best represents this sequence which allows predicting the imminence of the next degradation state and thus the defect of the studied system. In the second method (evolutionary Markov model), the computation of the probability that a sequence of observations arrives at a degradation state at the moment $t+1$, given the HMM modeled from the same sequence of observations, also allows us to predict the imminence of a defect. The third method predicts the imminence of a fault not by modeling the degradation process of the system, but by modeling each degradation state.

Chapter 3, Volume 1, "Electrical Signatures Analysis for Condition Monitoring of Gears in Complex Electromechanical Systems," written by S. Hedayati Kia and M. Hoseintabar Marzebali, deals with a review of their most remarkable research, which has been carried out in the last 10 years. A particular emphasis has been placed on the topic of noninvasive fault detection in gears using electrical signatures analysis. The main aim is to utilize the electrical machine as a sensor for the identification of gear defects. In this regard, a universal approach is developed for the first time by the authors which allows evaluating the efficacy of noninvasive techniques in the diagnosis of torsional vibration induced by the faulty gear located within the drive train. This technique can be considered an upstream phase for studying the feasibility of gear fault detection using noninvasive measurement in any complex electromechanical system.

Chapter 4, Volume 1, "Modal Decomposition for Bearing Fault Detection", by Y. Amirat, Z. Elbouchikri, C. Delpha, M. Benbouzid and D. Diallo, deals with induction machine bearing faults detection based on modal decomposition approaches combined to a statistical tool. In particular, a comparative study of a notch filter based on modal decomposition, through an ensemble empirical mode decomposition and a variational mode decomposition, is proposed. The validation of these two approaches is based on simulations and experiments. The achieved simulation and experimental results clearly show that, in terms of fault detection criterion, the variational mode decomposition outperforms the ensemble empirical mode decomposition.

Chapter 5, Volume 1, "Methods for Lifespan Modeling in Electrical Engineering", by A. Picot, M. Chabert and P. Maussion, deals with the statistical methods for electrical device lifespan modeling from small-sized training sets. Reliability has become an important issue in electrical engineering because the most critical industries, such as urban transports, energy, aeronautics or space, are moving toward more electrical-based systems to replace mechanical- and pneumatic-based ones. In this framework, increasing constraints such as voltage and operating frequencies enhance the risk of degradation, particularly due to partial discharges (PDs) in the electrical machine insulation systems. This chapter focuses on different methods to model the lifespan of electrical devices under accelerated stresses. First, parametric methods such as design of experiments (DoE) and surface responses (SR) are suggested. Although these methods require different experiments to organize in a certain way, they reduce the experimental cost. In the case of nonorganized experiments, multilinear regression can help estimate the lifespan. In the second part, the nonparametric regression tree method is presented and discussed, resulting in the proposal of a new hybrid methodology that takes advantages of both parametric and nonparametric modeling. For illustration purpose, these different methods are evaluated on experimental data from insulation materials and organic light-emitting diodes.

Chapter 1, Volume 2, "Diagnosis of Electrical Machines by External Field Measurement", by R. Pusca, E. Lefevre, D. Mercier, R. Romary and M. Irhoumah, presents a diagnostic method that exploits the information delivered by external flux sensors placed in the vicinity of rotating electrical machines in order to detect a stator inter-turn short circuit. The external magnetic field measured by the flux sensors originates from the airgap flux density and from the end winding currents, attenuated by the magnetic parts of the machine. In the faulty case, an internal magnetic dissymmetry occurs, which can be found again in the external magnetic field. Sensitive harmonics are extracted from the signals delivered by a pair of flux sensors placed at 180° from each other around the machine, and the data obtained for several sensor positions are analyzed by fusion techniques using the belief function theory. The diagnosis method is applied on induction and synchronous machines with artificial stator faults. It is shown that the probability of detecting the fault using the proposed fusion technique on various series of measurements is high.

Chapter 2, Volume 2, "Signal Processing Techniques for Transient Fault Diagnosis", by J.A. Daviu and R.A.O. Rios, revises the most relevant signal processing tools employed for condition monitoring of electric motors. First, the importance of the predictive maintenance area of the electric motors due to the extensive use of these machines in many industrial applications is pointed out. In this context, the most important predictive maintenance techniques are revised, showing the advantages such as the simplicity, remote monitoring capability and broad fault coverage of motor current analysis methods. In this regard, two basic approaches based on current analysis are explained: the classical methods, relying on the Fourier transform of steady-state current (motor current signature analysis – MCSA), and novel methods based on the analysis of startup currents (advanced transient current signature analysis – ATCSA). In the chapter, the most significant signal processing tools employed for MCSA and ATCSA are explained and revised. For MCSA, the basic problems derived from the application of the Fourier transform as well as other constraints of the methodology are explained. For ATCSA, the most suitable signal processing techniques are described, classifying them into continuous and discrete transforms. One representative of each group is accurately described (the discrete wavelet transform for discrete tools and the Hilbert-Huang transform for continuous tools), accompanying the explanation with illustrative examples. Finally, we discussed several examples of the application of each tool to electric motor fault diagnosis.

Chapter 3, Volume 2, "Accurate Stator Fault Detection in an Induction Motor Using the Symmetrical Current Components", by M. Bouzid and G. Champenois, deals with the accurate detection of stator faults such as inter-turns short circuit, phase-to-phase and phase-to-ground faults of the induction motor, using the symmetrical current components. The detection method is based on the monitoring of the behavior of the negative and zero sequence stator currents of the machine. This chapter also develops analytical expressions of these components obtained using the coupled inductance model of the machine. However, despite its efficiency, the negative sequence current-based method has its own limitations to detect accurate incipient stator faults in an induction motor. This limit can be explained by the fact that the negative sequence current generated in a faulty motor does represent not only the asymmetry introduced by the fault, but also by other superposed asymmetries, such as the voltage imbalance, the inherent asymmetry in the machine and the inaccuracy of the sensors. This aspect can generate false alarm and make the achievement of accurate incipient stator fault detection very difficult. Thus, to increase the accuracy of the fault detection and the sensitivity of the negative sequence current under different disturbances, this chapter proposes an efficient method able to compensate the effect of the different considered disturbances using experimental techniques having the originality to isolate the negative sequence current of each disturbance. The efficiency of all these proposed methods is validated experimentally on a 1.1-kW motor under different stator faults. Moreover, an original monitoring system, based on neural networks, is also presented and described to automatically detect and diagnose incipient stator faults.

Chapter 4, Volume 2, "Bearing Fault Diagnosis in Rotating Machines", by C. Delpha, D. Diallo, J. Harmouche, M. Benbouzid, Y. Amirat and E. Elbouchikhi, is focused on detection, estimation and diagnosis of mechanical faults in electrical machines. Nowadays, it is necessary to rapidly assess the structural health of a system without disassembling its elements. For this *in situ* diagnosis purpose, the use of experimental data is very imperative. Moreover, the monitoring and maintenance costs must be reduced while ensuring satisfactory security performances. In this chapter, we focus on vibration-based signals combined with statistical techniques for bearing fault evaluation. Based on a four-step diagnosis process (modeling, preprocessing, feature extraction and feature analysis), the combination of several techniques such as principal components analysis and linear discriminant analysis in a global approach is explored to monitor the condition of vibration-based bearings. The main advantage of this approach is that prior knowledge on the bearing characteristics is not required. A particularly reduced frequency analysis has led to efficiently differentiate the bearing fault types and evaluate the bearing fault severities.

Chapter 5, Volume 2, “Diagnosis and Prognosis of Proton Exchange Membrane Fuel Cells”, by Z. Li, Z. Zheng and F. Gao, deals with the diagnostic and prognostic issues of fuel cell systems, especially the proton exchange membrane (PEMFC) type. First, the basic functioning principle of PEMFCs and their current development and application status are presented. Their high cost, low reliability and durability make them unfit for commercialization. In the following sections, degradation mechanisms related to both the aging effect and the system operations are analyzed. In addition, typical variables and characterization tools, such as polarization curve, electrochemical impedance spectroscopy, linear sweep voltammetry and cyclic voltammetry, are introduced for the evaluation of PEMFC degradation. Various diagnostic and prognostic methods in the literature are further classified based on their input-to-output process model of the system, namely model-based, data-driven and hybrid methods. Finally, two case studies for diagnosis and prognosis are given at the end of each part to give the readers a general and clearer illustration of these two issues.

1

Diagnostic Methods for the Health Monitoring of Gearboxes

1.1. Introduction

Maintaining the safety and availability of electromechanical systems (machines) while minimizing operating costs is one of the major challenges industry faces today. To achieve this, machinery health monitoring becomes a necessity. The choice of monitoring techniques depends on the nature of installation and potential degradation. These are usually processes that can either be performed during operation (e.g., monitoring temperature) or delayed (e.g., lubricant analysis), thus requiring scheduled shutdowns of the machines. Moreover, upon diagnosis and depending on the reliability level of the machine operation, a maintenance process can be implemented. The first form of maintenance is called "corrective maintenance." This strategy is followed only after the appearance of a failure. This strategy involves sudden and uncontrolled shutdowns. The second form of maintenance is called "preventive maintenance." This is conducted at predetermined intervals or based on a certain pre-established criterion. Hence, preventive maintenance includes two types of maintenance strategies: systematic and conditional. Systematic preventive maintenance is scheduled periodically without taking into account the machine diagnostic operation, and there is no intervention before a predetermined deadline.

It is based on different criteria, for example, the number of hours of operation or the number of cycles of use. Maintenance operations such as lubrication or oil changes are examples of preventive maintenance. These are

---

Chapter written by Abdenour SOUALHI and Hubert RAZIK.

not curative interventions but actions, usually simple and inexpensive, seeking to minimize the failure rate. Nevertheless, following the geometry dispersions of components, the heterogeneity of materials used and the possible surcharges of operations, degradation may occur before the scheduled deadline. In addition, a defect does not appear during the expected period of time, leading to the replacement of components in a good condition which in turn leads to added maintenance cost. Thus, conditional preventive maintenance must be adopted when reliability is the key factor. This strategy is based on features characterizing the health status of the machine to predict and diagnose the appearance of possible failures before they reach the stage causing the breakdown of the machine. Hence, this chapter details and contextualizes the diagnosis of gears. It also presents failures that appear in gearboxes comprising gears and their types. Then, it introduces analysis methods that are generally used, followed by the presentation of main monitoring tools. Finally, we present the state-of-the-art of various techniques for diagnosing gear defects.

1.2. Identification of critical components in gearboxes

Figure 1.1 shows a gearbox with a set of interconnected mechanical components integrated within a housing. The gearbox mainly comprises a gear train that performs the reduction of motion connected to an electric motor that acts as a power generator. Bearings are also part of the gearbox, to guide rotation. They are connected to rotating shafts, thus serving as a supporting structure for the gearbox. Finally, the last elements of the gearbox are the driven mechanisms, including the couplings that ensure transmission of torque, mechanical load output or brakes that allow the slowing down, or even immobilization, of the movement of the assembly. All of the aforementioned elements have influence on the dynamic behavior of the gearbox, but it is generally accepted that the gear train is one of the main sources of excitation, and gears also respond very well to the specific performance, efficiency and power requirements imposed by modern standards. However, conditions such as comfort, vibratory behavior and reliability impose new technological and economic pressure on this component. At the same time, bearings are also subjected to relatively important functional constraints.

Figure 1.2 shows the results of various researches studying origins of faults and their location in a power transmission-based gearbox and also demonstrate that gears are the most sensitive component [BRE 02]. Bearings come in second. Thus, in addition to manufacturing and/or assembly defects, there are numerous causes that lead to damage and hence give rise to anomalies with several degrees of severity. So, from both industrial and scientific points of view, it is important to focus gearbox diagnostic efforts primarily on gears and secondarily on bearings.

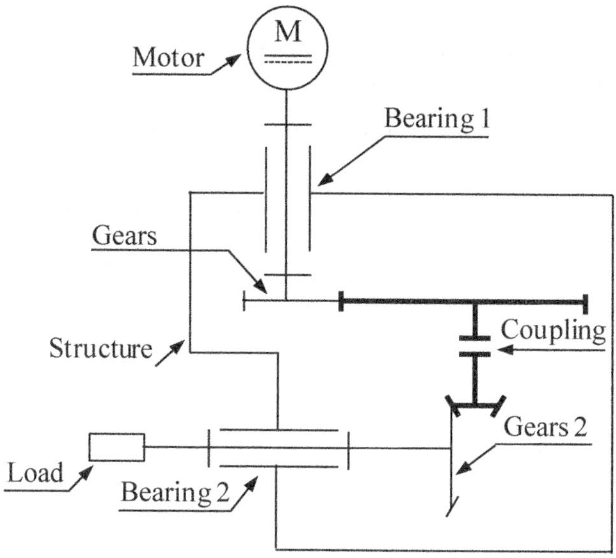

Figure 1.1. *Example kinematic diagram of a gearbox*

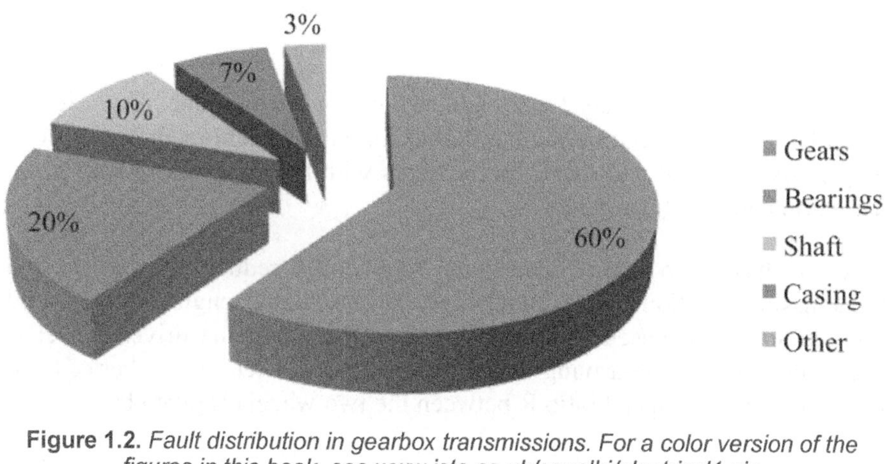

Figure 1.2. *Fault distribution in gearbox transmissions. For a color version of the figures in this book, see www.iste.co.uk/soualhi/electrical1.zip*

1.2.1. *Types of gears*

A gear mechanism contains two gears – driving and driven gears – meshing with each other. This association ensures the transmission of movement, and hence mechanical energy, between two shafts with a constant speed [MS 85].

Figure 1.3 shows the following classes of gears, depending on the relative position of the two connected shafts:

– Parallel gears, when the alignment of shafts is parallel (the planetary gears, for example, are part of it);

– Bevel gears, when there is an intersection of shaft extensions;

– Worm gears, when shafts are in any relative position (wheels and worms, for example, are part of it).

(a) (b) (c)

Figure 1.3. *(a) Parallel gears, (b) bevel gears and (c) worm gears*

The teeth of gears may be straight, helical or herringbone. Thus, forces are either (i) only radial on the parallel gears with straight teeth or chevrons or (ii) mixed (radial and axial) on the gears with helical gears as well as on the worm and bevel gears.

Gears have two major operations: they either reduce or multiply the transmission rotation speed and hence the associated engine torque. The rotation speed increases when the number of teeth of the driven wheel is higher than that of the driving wheel and decreases when its number of teeth is lower. Thus, the speed ratio R between the two wheels is given by

$$R=\frac{Z_1}{Z_2}, \quad [1.1]$$

where Z_1 and Z_2 are the number of teeth of the driven and driving wheels, respectively.

Consider a gear mechanism grouping two gears with the number of teeth Z_1 and Z_2 rotating at frequencies f_1 and f_2, respectively. When a tooth of the driving wheel meshes with that of the driven wheel, a periodic load is generated at the meshing rate of the teeth [AUG 02]. This shock occurs according to the mesh frequency f_e, which is equal to the rotation frequency of the considered wheel multiplied by its number of teeth, as shown in the following equation:

$$f_e = f_1 Z_1 = f_2 Z_2. \qquad [1.2]$$

Owing to the change in rigidity, the meshing process causes a sliding of the teeth that produces vibrations and torque variations. The majority of natural phenomena are nonlinear, and most of these signals have variable frequency. Thus, a defect in a tooth changes the amplitude and phase modulations of the dynamics of the gear, which will be analyzed specifically for possible faults.

1.2.2. *Types of gear defects*

Gears are generally subjected to high torque and severe operating conditions, thus causing failures that must be monitored. For this purpose, it is imperative to understand the different types of defects encountered [BAD 99]. The main damages occurring in gears are: (i) degradation distributed over all the teeth and (ii) degradation located on a part of the teeth. They appear mainly during the transmission. Manufacturing damages, such as the error of tooth profiles and wheel eccentricity, and installation faults, such as parallelism defects, are not discussed in this chapter.

1.2.2.1. *Distributed faults*

(a) (b)

Figure 1.4. *(a) Wear and (b) pitting*

1.2.2.1.1. Wear

Wear is a phenomenon of material loss due to the abrasive friction of two surfaces in contact with each other, as shown in Figure 1.4(a). This phenomenon is thus caused by repeated surface stresses, of which its importance exceeds the permissible endurance of the material. This type of wear increases with increasing sliding speed and hence the resulting mechanical load, as well as with the influence of the abrasive elements introduced into the lubricating product. The progression of natural wear is inversely proportional to the surface hardness of the tooth and propagates on the surface continuously. This wear occurs when the lubricant contains abrasive particles or when it has a corrosive nature. The lack of wear can lead to faulty operation as a result of the abnormal change in friction, which can hinder the movement.

1.2.2.1.2. Pitting

Pitting are shallow cavities, not exceeding 0.3–0.5 mm in depth, which may affect all teeth of the gear, as shown in Figure 1.4(b). This type of defect particularly affects the areas of the pitch circle of the teeth. This deterioration generally takes place in less-hard steel construction gears and, therefore, is more sensitive to cumulative effects of the overloads. When the viscosity of the lubricant is sufficiently high, the oil film separating the contact surfaces is thick and, therefore, the risk of occurrence of pitting is less. Thus, the existence of pits is related to a ratio of the lubricating film thickness to the composite roughness of the material, which is too insufficient to avoid the contact between the irregularities of the surfaces. This failure occurs as a result of misalignments and deviations of axes.

1.2.2.2. *Localized defects*

(a) (b) (c)

Figure 1.5. *(a) Chipping, (b) cracking and (c) scuffing*

1.2.2.2.1. Chipping

Chipping is a phenomenon appearing in the form of deep and wide holes but is less common compared to pitting, as shown in Figure 1.5(a). The failure occurs under the effect of undercoating fatigue, particularly at the maximum shear point of the gear. It is caused by the effect of the surface pressure of the contact when it exceeds the permissible tolerance threshold of the material. This overshoot is often attributed to the poor mechanical load distribution. Chipping usually occurs in cemented gears which have a superficial hardened surface layer and are therefore not affected by the wear phenomenon. These gears are widely used for their ability to transmit considerable mechanical torque while being characterized by reduced dimensions. The spread of chipping can quickly cause tooth breakage, which is considered a consequence of this phenomenon rather than a defect, thus causing disengagement of the meshing operation.

1.2.2.2.2. Cracking

Cracking occurs under the effect of fatigue mainly in charging cycles, as shown in Figure 1.5(b). It occurs in areas such as in the rounded toe of the teeth or the edges of the teeth solicited in traction, where stress is important. The manifestation of this phenomenon is caused by exceeding the stresses applied to the toe of the teeth compared to the limit of fatigue and the elastic limits of the gears. The induced cracking weakens the tooth and reduces the rigidity of the material, thus promoting the propagation of cracks on its entire surface. This degradation usually occurs in fine and hardened steels which prove to be particularly sensitive to the concentration of traction stresses at the level of the teeth.

1.2.2.2.3. Scuffing

Scuffing is the result of the destruction of the lubricant-protecting film under the effect of friction under load, and this degradation is generated by the increase in temperature (see Figure 1.5(c). This damage is related to the metal–metal contact occurring during meshing and leads to the removal of the metallic material. The probability of the occurrence of this scuffing phenomenon depends primarily on the condition of the lubricant and the service conditions. Scuffing is confined to the sliding zones and increases the deformation of the profiles of the teeth. This deformation induces anomalous variations in angular velocities and pressure angles.

1.3. Diagnostic methodology

In the context of the conditional preventive maintenance strategy, the detection of a defect must be part of a well-defined diagnostic methodology in order to optimize the maintenance process. This methodology is part of the supervision which is the set of processes needed to respond correctly to abnormal changes in the state of a system. These changes occur in the form of failures, malfunctions or untimely stops of the operation. Thus, the supervision process helps make the right decisions in terms of scheduled maintenance actions in a timely manner to maximize reliability.

Figure 1.6 presents the three main tasks associated with the general concept of diagnosis in the context of supervision [AYA 15]:

– Analysis. It consists of a set of technical and physical processes that ensure the control of magnitudes representative of good or bad functioning of the component studied. It is, therefore, necessary to monitor the state of health of the elements of the system based on their measured response in operation.

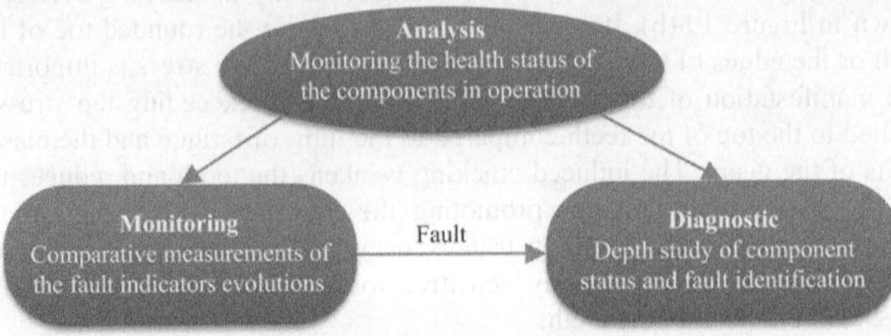

Figure 1.6. *Diagnostic methodology*

– Monitoring. It is the action of evaluating the state of the process during operation. For example, monitoring uses knowledge and information about the condition in question, including defect indicators. This step then detects the symptoms indicating a possible abnormal change and characterizes it.

– Diagnosis. It is the identification and location of the detected defect. This knowledge serves to characterize the severity emanating from the defect and guides the action to follow accordingly. It exploits the symptoms detected in the monitoring phase to confirm or deny the presence of defects and categorize them before the decision is made.

1.4. Methods of analysis

As discussed above, analysis techniques are methods that monitor the operation of machines as part of the conditional preventive maintenance. The most commonly used techniques for diagnosing rotating machines are temperature analysis, oil analysis, acoustic analysis and, especially, vibration analysis. Recently, new analysis techniques, including electromagnetic torque analysis, flow analysis and electrical current analysis, are receiving increased interest. These techniques are discussed in detail in the following sections.

1.4.1. *Temperature analysis*

Temperature analysis consists of controlling the temperature at the neighborhood or surface of a given material. It is based on the use of temperature probes or infrared thermography. The result of the latter is a two-dimensional cartographic measurement, called a "thermogram." The infrared radiation acquired is mainly emitted by the observed surface, and after prior calibration, it can be converted into a measurement of the temperature. This analysis detects most of the defects involved in heating of the material, such as lubrication problems. The interpretation of the results of the temperature control is often fast, however the analysis does not allow for a thorough diagnosis. In addition, monitoring is limited by the visibility of the outputs of the sensors used (infrared cameras) and it is generally difficult to detect early defects [LEE 08].

1.4.2. *Oil analysis*

The analysis of the oil is susceptible to give an indication on the state of the oil, especially on the state of systems soaked in oil baths such as gears or bearings. Several factors can be monitored, such as the degradation of the oil over time following the oxidation which is often detected by a spectrographic analysis. Also, the physicochemical properties, such as the viscosity of the oil and the water content, indicate an increase in temperature of the element concerned. Also mentioned is the chip detector that determines the presence of metal particles polluting the oil and characterizing the wear of the component. However, it also has limitations such as the inability to precisely locate the defect and the necessary precautions to be taken to carry out the sampling [ARQ 09].

1.4.3. *Acoustic analysis*

Acoustic analysis is based on the study of sounds in low or high frequency via acoustic emission based on ultrasound (sounds with frequency exceeding 20 kHz) to define the state of health of materials. This technique is capable of detecting phenomena such as the propagation of cracks and friction within the rotating elements.

Numerous analogies exist between vibratory analysis and acoustic emission. Acoustic emission is caused by static or dynamic deformation within a mechanical element under stress. The deformation creates internal micro-displacements, thus producing high-frequency waves. The rapid attenuation of these waves forces the acoustic sensor to be placed near the source. This constraint introduces sensitivity errors, especially with regard to the ambient noise. Thus, this analysis is often difficult to implement and presents problems inherent in the repeatability of the measurements [MIG 11].

1.4.4. *Vibration analysis*

Vibration analysis is the most commonly used analysis technique in industries. The study of the defects of rotating machines via vibration signals ensures the detection of a wide range of defects, including mechanical. Indeed, it is possible to detect kinematic defects such as eccentricities and misalignments, as well as structural defects such as wear and cracks. Vibratory analysis retranscribes the kinematic behavior generated by a structure solicited by a dynamic excitation. Acceleration is the magnitude most often measured using accelerometers placed on structures in radial, axial and vertical directions. Also, velocity and displacement, which are the mathematical integrals of acceleration and which can be measured using encoders and position sensors, fall within the scope of vibration analysis [SHKB 95].

This analysis has the advantage of detecting failures at an early stage and, therefore, offers the possibility of a thorough diagnosis. However, it also has several limits which stem mainly from the technical constraints of the sensors. It is necessary to place the sensor as close as possible to the process due to the sensitivity problems. Thus, this constraint generates an accessibility problem as well as difficulties in the installation of the sensors in order to make acquisitions in a neighborhood sufficiently close to the defect. In addition, vibration analysis also involves considerable economic constraints in view of the implementation costs and sensor prices significantly higher than other techniques. Hence, research increasingly tends to focus on other methods that may complement or even replace vibration analysis [CAS 03].

1.4.5. *Electromagnetic torque analysis*

Electromagnetic torque analysis is based on the detection of faults through the search for harmonics in the frequency domain. These harmonics come from an interaction between the flux and the current. Thus, reconstruction of this torque from the measurements of the electric machine or by using a physical model representing the system is possible. Use of this analysis may prove to be a wise choice for the detection of defects in the mechanical load. Indeed, fluctuations of the load torque have repercussions on the flows and the currents of the machine, causing their variations. As a result, when anomalies occur in the induction machine, such as misalignment of the shaft or degradation of the bearing, the electromagnetic torque generates oscillations, resulting in the appearance of harmonics [HTS 15].

1.4.6. *Flux analysis*

Flux analysis is based on the principle that imbalances, whether mechanical or electromagnetic, can have an impact on the distribution of the electromagnetic field as well as on the electromechanical conversion of the drive motor. In this context, several studies have been carried out to implement the extraction of flux measurement at the air gap of the motor and also the axial and leakage flux. For these measurements, exploratory coils installed on the cast iron of the motor are used. Thus, leak flow analysis, for example, identifies electrical faults such as short circuit in the induction motor. Also, some components of the axial flow ensure the detection of defects related to bearings and rupture of rotor bars. The main limitation of this process is its high dependence on the resulting load, as the best sensitivity is obtained at full load [CAS 03].

1.4.7. *Analysis of electric currents*

The current analysis of electrical motors is a promising method for monitoring a wide range of mechanical faults, particularly phenomena involving torque or rotational speed fluctuations, in addition to failures of an electrical nature. The electric current signal of the electric motor is sensitive to the modulations induced by dynamic excitations, in particular, torque oscillations. From there, it is possible to quantify the frequencies and forms of the modulations in order to locate the origin of the defect. This technique is based on amperometric clamps and transformers for a measure that has the advantage of being noninvasive and inexpensive [KHC 15].

The global applications of current analysis in the scientific and industrial fields can be classified into four main areas [VPM 15]:

– The most common area is equipment diagnostics and power grids. The majority of these installations have already been controlled for several years in terms of usual electrical measurements, such as insulation and supply controls. Recently, the novelty consists of adding to the classical systematic controls, the conditional analysis of the modulations of the electric currents in order to characterize locally the functioning of the elements of the system.

– Current analysis is also used in the study of shaft defects. These failures occur in the form of clearance in clutches, keyways or in torsional resonances and introduce abnormal fluctuations in rotational speed and mechanical torque.

– The analysis of defects in manufacturing processes is also an area of application of current analysis. Indeed, manufacturing errors generate problems of acyclism or regulation at the level of the transmission, inducing relatively large modulations in the electrical signal.

– The last field of application represents the control of alternative machines, such as heat engines or alternative pumps, where acyclism is inherent to the operation. The influence of the acyclic phenomenon on the driving motor current induces modulations that allow for the monitoring of the operation.

1.5. Monitoring methods

Monitoring a machine consists of tracking the evolution of the defect before it causes an untimely and unplanned shutdown. Thus, it maximizes reliability by reducing downtime and improving the life of the system. Monitoring ensures control of the behavior of the machine by comparing the successive readings of the signals collected during the analysis phase. A possible malfunction of the system is signaled, as soon as there is a noticeable change in the given indicators compared to the reference values.

In this context, the notion of a fault signature is introduced, which is defined as the quantized reference that indicates the occurrence of a given defect by comparison with a detection threshold representing its ability to detect the defect concerned. When the indicator changes during operation, at least two thresholds can be defined: (i) an alarm threshold which detects the presence of a fault and (ii) a danger threshold which indicates an impending failure caused by the defect in question (see Figure 1.7). Generally, the signature of the defect is established following a first measurement campaign in the system in the healthy state which constitutes the reference operation [AYA 15].

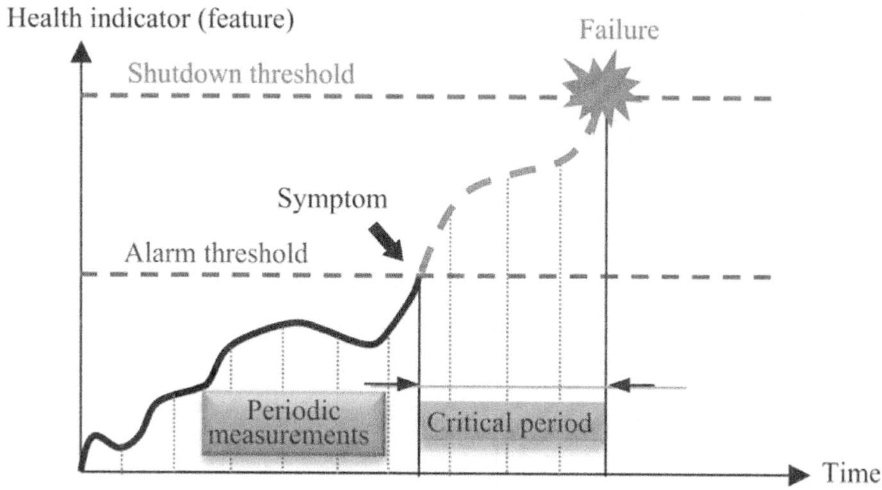

Figure 1.7. *Notion of a health indicator [AYA 15]*

Monitoring is based on the processing of the signal acquired previously during the analysis step. The signal processing is based on a set of techniques dedicated to the study and the transformation of the input signals in order to carry out the extraction of the characteristics of the defects, in particular, the signatures. In fact, the information contained in the measures is often not directly accessible and requires an intermediate processing step before it can be used. Several monitoring techniques have been used in the literature [PCE 13, CLW 12, TMMZ 12, NY 10]. Thus, it is possible to classify these techniques into three domains: (i) time analysis, (ii) frequency analysis and (iii) time–frequency analysis. The most frequently used techniques for gear monitoring are presented in this chapter.

1.5.1. *Temporal analysis*

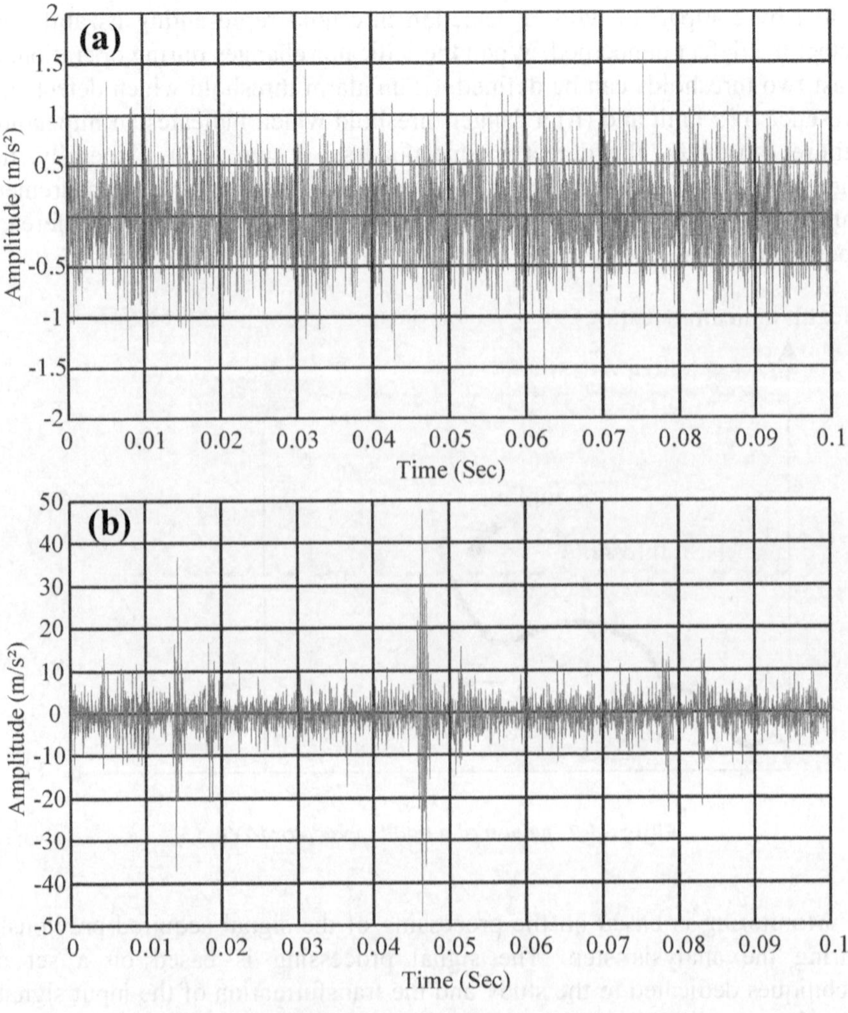

Figure 1.8. *Waveform of a temporal vibration signal: (a) healthy and (b) with defect*

Conventional time analysis aims at exploiting the temporal information as well as the information relative to the amplitudes contained in the signals to detect the defects. The variation in the amplitude of the signal reflects the presence of the fault, and the periodicity indicates the possible source of the fault [LOU 04]. Time domain approaches are appropriate when periodic components are observed and when faults generate broad frequency bands following periodic pulses. As shown in Figure 1.8, time waveforms show changes in fault signatures, but it is often difficult to diagnose the source of failures. Indeed, some mechanical systems generate significant torque variations during their operation. When a progressive fault arises, the level of variation in the torque is likely to increase steadily over time, but this increase may be difficult to assess [HLC 00].

Mechanical processes are said to be deterministic if their properties, such as displacement or acceleration, are predicted over time. In addition to these deterministic components, mechanical systems such as faulty gears have random or nondeterministic characteristics that are difficult to predict but can nevertheless be estimated using statistical indicators. These indicators are called "statistics," as they are based on the analysis of the temporal characteristics of registered signals [TCZ 15]. The most commonly used indicators are: the root mean square (RMS) used to measure the mean energy of a signal, the peak-to-peak value used to quantify the maximum amplitude between the extreme values of the signal, the kurtosis used to quantify the impulsiveness of the signal, and so on (see Table 1.1). The basic reason behind using these indicators is that any appearance of a defect causes a significant change in the statistical characteristics of the signal and therefore on the features.

Peak value	$\hat{x} = \sup_{i=1,\dots,N} \lvert x_i \rvert$
Average value	$\overline{x} = \frac{1}{N}\sum_{i=1}^{N} x_i$
Standard deviation	$\sigma = \sqrt{\frac{1}{N-1}\sum_{i=1}^{N}(x_i - \overline{x})^2}$
RMS	$\text{RMS} = \sqrt{\frac{1}{N}\sum_{i=1}^{N}(x_i)^2}$
Skewness	$\text{SK} = \frac{\sum_{i=1}^{N}(x_i - \overline{x})^3}{(N-1)\sigma^3}$
Crest factor	$\text{CI} = \frac{\hat{x}}{\sqrt{(1/N)\sum_{i=1}^{N}(x_i)^2}}$
Kurtosis	$\text{KU} = \frac{\sum_{i=1}^{N}(x_i - \overline{x})^4}{(N-1)\sigma^4}$

Table 1.1. *Statistical features*

For example, the peak-to-peak value and the kurtosis are sensitive to the shock-type defects, whereas the RMS value is sensitive to energy dissipations. The advantages of temporal analysis are its simplicity of implementation and its interpretability. This technique is, therefore, particularly well suited for online monitoring. Nevertheless, these features often generate "false alarms" when the signals are not Gaussian.

Here x_i is the *i*th sample of the recorded signal $x(t)$ and N is the number of samples.

The most popular indicators used in the literature are the following:

– The standard deviation σ is a measure of the signal's dispersion about the mean. It should be noted that the standard deviation is similar to the RMS. This feature is sensitive to the energy dissipations and used as a metric in classifiers such as dynamic Bayesian networks [BS 95] and neural networks [LEE 13].

– The RMS value is the second statistical moment of the signal. It is the most interesting measure of vibration amplitudes. In addition to taking into account the evolution of the signal over time, the calculation of the RMS value is related to the vibratory energy and therefore to the "potential for deterioration" of the vibration. Recently, the RMS was used to verify the effectiveness of exploiting only a selection of the vibration signals instead of the original ones. Results showed that when RMS is applied after the selection step, it provides more significant information about the faulty cases [FWZ +16].

– The peak-to-peak value indicates the difference between the maximum and minimum magnitudes of the movement. It is a useful quantity to determine the displacement of a component, of which displacement is critical for considerations of maximum stress or mechanical slack.

– The peak value gives the maximum amplitude and is useful in measurements concerning phenomena of short duration (e.g., shocks). However, it does not take into account the evolution of the vibration over time.

– The average value takes into account this evolution but, unlike the RMS value, it has no relation to any physical quantity.

– The crest factor is defined as the ratio of the peak value to the RMS value of the signal. A localized fault generates a strong peak amplitude vibration and low effective amplitude, thus generating a large crest factor. For generalized faults, the vibrations generated have strong peak and effective amplitudes, and therefore a low crest factor. The crest factor remains low (between 2 and 6). A localized defect generates a high peak amplitude and low RMS amplitude, and so an important peak factor (greater than 6). However, as the RMS value increases for progressive failure, the crest indicator decreases [DBLR 04]. The major inconvenience of this indicator is that it presents approximately the same values in the good state and in the end of the bearing life.

– Besides these basic quantities (RMS value, peak value and crest factor), kurtosis is another more complex indicator. This is related to the shape of the distribution representing the noise generated by a bearing. Normally, the statistical distribution of this variable obeys Gaussian law. However, when a fault appears, the shape of the distribution changes. Kurtosis (or statistical moment of order 4) then allows us to characterize "flattening."

Thomas [THO 02] used kurtosis to establish a criterion of severity for the diagnosis of machines, as shown in Table 1.2. As it is calculated on several frequency bands starting from just a few hertz and the first defects are manifested by low-frequency shocks, kurtosis can detect the degradation of bearings at an early stage. In addition, it is suitable for monitoring rotating shaft bearings at low rotational speeds (less than 600 rpm).

This method also has the same drawback as the crest factor. Kurtosis decreases when the defects develop. At an advanced degradation state of the bearing, the kurtosis value is close to 3, representing a state without fault.

Kurtosis	Severity
1.5	Good
2.8–3.2	Acceptable
3.2–4	High
>4	critical

Table 1.2. *Quantification of the severity according to kurtosis*

– Skewness is defined by the statistical moment of order 3 normalized by the cube of the standard deviation. It is considered as an asymmetry coefficient measuring the asymmetry distribution of a vibration signal compared to its mean value. This indicator gives information about the morphology of the signal's envelope. A positive skewness value corresponds to a distribution shifted to the lowest peaks (the distribution is shifted to the right compared to its mean), whereas a negative value corresponds to a distribution shifted to the highest peaks (the distribution is shifted to the left). If the skewness is zero, the distribution is symmetrical (Gaussian distribution, for example). It was proved in the paper of Hemmati *et al.* [HOG 16] that skewness is a good indicator for diagnosing bearing faults.

Table 1.3 summarizes a comparative study conducted by Dron *et al.* [DBLR 04] to observe the evolution of the RMS, kurtosis, skewness and crest factor in the case of degradation of a gearbox. The degradation test was done for a duration of 12 days. These indicators are shown in Figure 1.9. These curves show the beginning of the degradation after the 8th day followed by an increase until the apparition of the degradation on the 12th day.

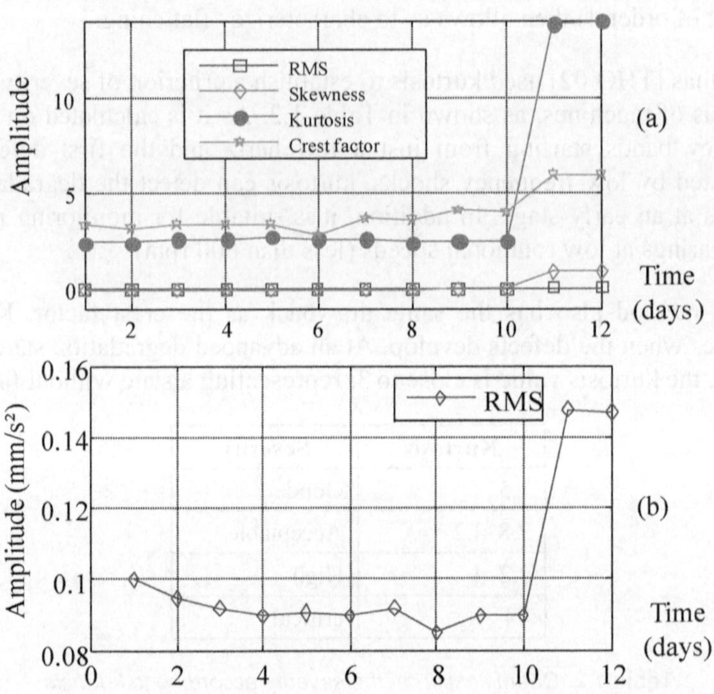

Figure 1.9. *(a) Evolution of the kurtosis, skewness, crest factor and the RMS; (b) zoom on the RMS curve*

Entropy, a noncommon indicator, is added to Table 3. The different types of entropy are listed in Table 1.4 and have been developed in [GY 03] to analyze the experimental data according to different flow regimes in a pipe.

The most interesting factor about entropy is that it varies with noise, especially the lower Shannon entropy which indicates that the signal is periodic and has a high signal-to-noise ratio.

Here n is the number of samples, x_i the ith element of samples and p the threshold.

	Main benefits	**Main limitations**
RMS, peak value, standard deviation, average value	– Simple and reliable indicator	– Late detection – Difficult diagnosis – Poorly adapted to low rotation speeds – Determination of empirical thresholds
Kurtosis	– Suitable for monitoring bearings of shafts rotating at low speeds (<600 rpm) – Early detection – High sensitivity to periodic and nonperiodic shocks	– Decrease in indicator at the end of the life of the bearing – Diagnosis often difficult
Crest factor	– Independent of the operating conditions (bearing dimensions, load, rotational speed)	– Decreases when defects develop

Table 1.3. *Comparative study of statistical features*

Entropy	**Definition**
Shannon entropy	$-\sum_i x_i^2 \log(x_i^2)$
Log energy entropy	$\sum_i \log(x_i^2)$
Threshold entropy	$\begin{cases} 1, & \text{if} \mid x_i \mid > p \\ 0, & \text{otherwise} \end{cases}$
Norm entropy	$\lvert x_i \rvert^p$

Table 1.4. *Types of entropy*

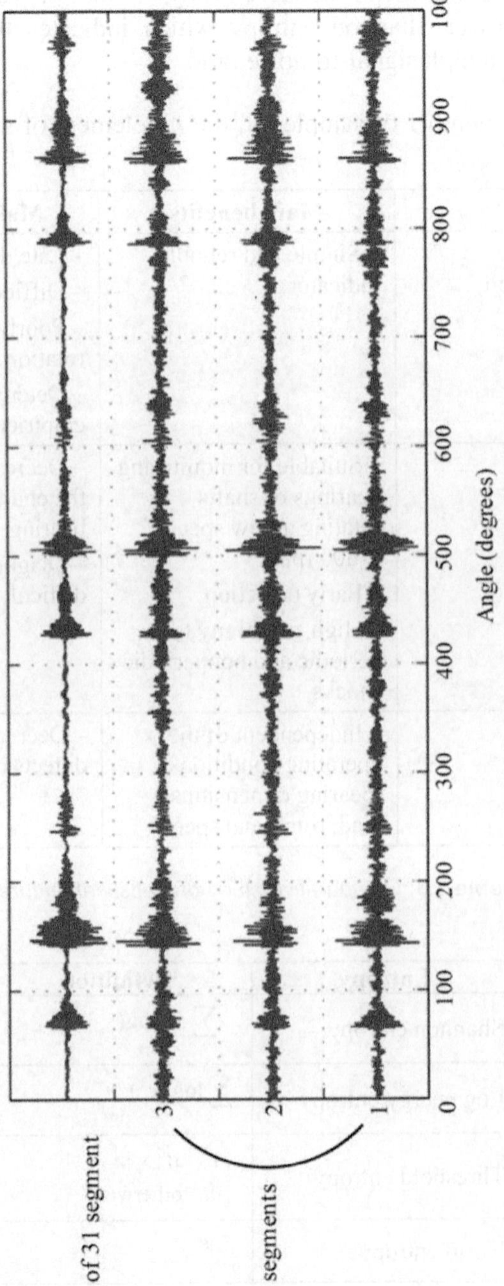

Figure 1.10. *TSA applied to a gearbox [HW 96]*

Another temporal analysis tool is the time synchronous average (TSA) used for cyclostationary signals introduced by Bennett in 1958 [TCZ 15]. This tool is used for fault detection in rotating equipment, especially in gearbox fault diagnosis due to the cyclostationarity of its vibration signals. It consists of dividing a vibration signal into segments and carrying out a mean on these segments. The length of each segment is taken to be equal, in the case of a gearbox, successively to the rotation period of the pinion and the gear wheel. TSA is used to eliminate the noise as a means of preprocessing the signal. Hammond and White [HW 96] showed that the angular synchronous average has more satisfactory results than TSA in fault isolation. For instance, the fault can be detected and isolated where, at 130°, the profile of the shock between the faulty pinion and the gear wheel is noticeable (see Figure 1.10).

Yang [YAN 04] proposed another temporal analysis to monitor gear faults. This analysis combines fast dynamic time warping (fast DTW) and correlated kurtosis (CK) techniques. This approach aims at detecting the local gear fault and determining the faulty gear and its position. Knowing that the faulty gear tooth generates periodic impulses in the vibration signal, fast DTW is used to extract them by using a reference signal at the same frequency of the nominal gear mesh harmonic. It is based on vibration signals obtained from a healthy and steady functioning of the system. Then the subsequent signal is resampled for subsequent diagnostic analysis using the CK technique, which aims at isolating the gearbox fault locally by exploiting the periodic effect of the fault. The proposed approach is given in Figure 1.11.

Hawwari and Soualhi proposed another temporal analysis with a new concept of passing the traditional one-dimensional representation to a two-dimensional representation of the vibration signal [HS 17]. This approach transforms a vibration signal into an image by translation, and the significant features are deduced by using the scale invariant feature transform (SIFT) in order to detect faults by affecting the concerned vibration signal to the corresponding faulty category (diagnosis level). For the translation (see Figure 1.12), the amplitude of the signal samples is normalized to have values between 0 and 255. Then these values are put in a matrix $M \times N$, where the coordinate of the ith element in the vibration signal is pixel (j, k) in the matrix with $j = \text{floor}(i/N)$ and $k = \text{modulo}(i/N)$. Then the SIFT algorithm is applied on this image to obtain 128 dimension vectors. Finally, each feature vector is compared to each centroid of fault category dictionary in order to obtain a histogram of the number of matching feature vectors to the category. The fault category of the vibration signal is that which has the highest number of match. Figure 1.12 gives an example of this method.

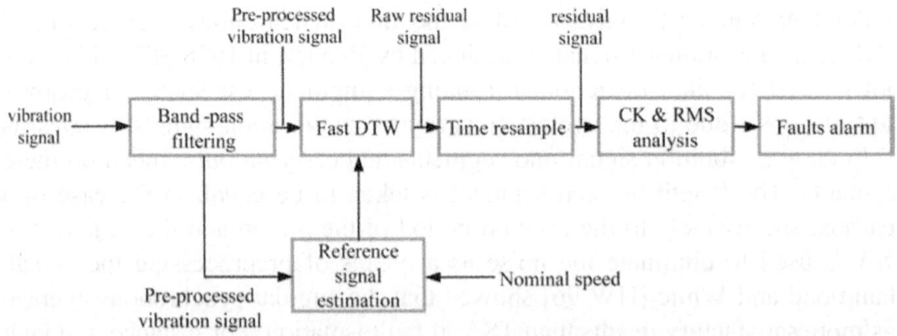

Figure 1.11. *Flowchart of the proposed time domain fault detection method [UA 82]*

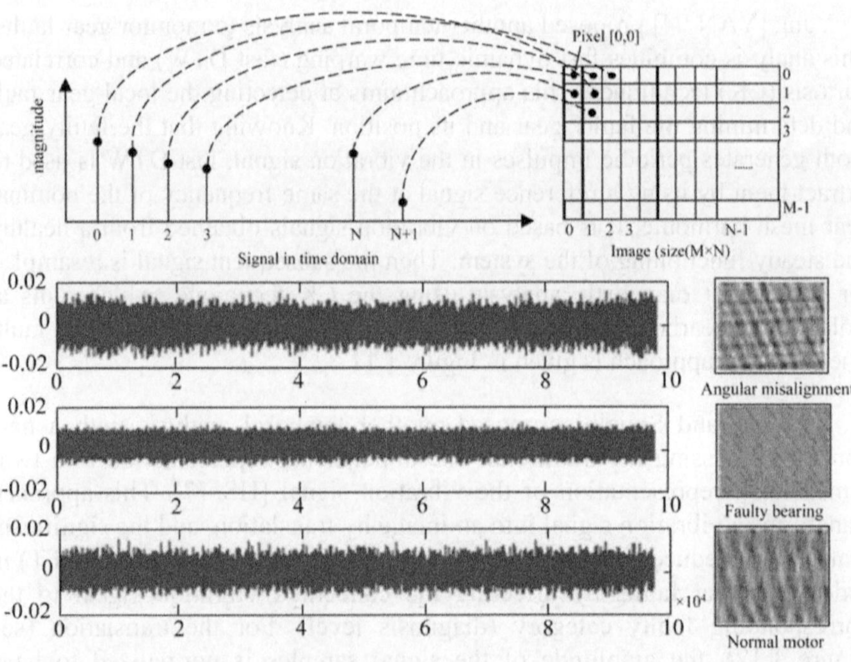

Figure 1.12. *An example of vibration signal translated into the 128–128 gray image (M = 128 and N = 128) [HS 17]*

It is well known that bearings generate modulations in the vibration signal corresponding to the characteristic frequency of flatting. Because of this phenomenon, the envelop analysis of vibration signals appears to be a powerful tool for the diagnosis of bearings. It enables the measurements to be linearized in the case of weak nonlinearities, which is the case in the presence of bearing defects. The objective of the envelope analysis is to dissociate the effects of gear defects from those of bearing defects as well as to identify the type of bearing defects. Vibration signals are first windowed (Blackman window) and then filtered using a bandpass filter in which the range mainly focuses on bearing fault frequencies. The envelope of the signal is the Hilbert transform of the filtered windowed signal. Thus, the obtained envelope does not provide an indication of the presence of a defect. The diagnosis is established from the spectral or cepstral representation of the envelope.

Figure 1.13 highlights the signature of the inner ring fault located on the roller bearing. The presence of the inner race of the roller bearing produces in the spectrum an emergence of the first two harmonics and some lateral bands around these peaks (Figure 1.13(a) and (b)). The spacing between the secondary peaks is the rotation frequency of the motor shaft which carries the defect.

The envelope analysis of vibration signal is very insensitive in the case of gear defects. Figure 1.14 shows the spectral analysis in the case of a local defect on a tooth. No characteristic peak emerges.

In the case of combined faults (gears and bearings), gear damage does not interfere with the diagnosis of the bearing faults (see Figure 1.15). The spectrum of the vibration signal in the case of ball bearing defect alone and in the case of combined ball bearing and spalling gear defects are the same.

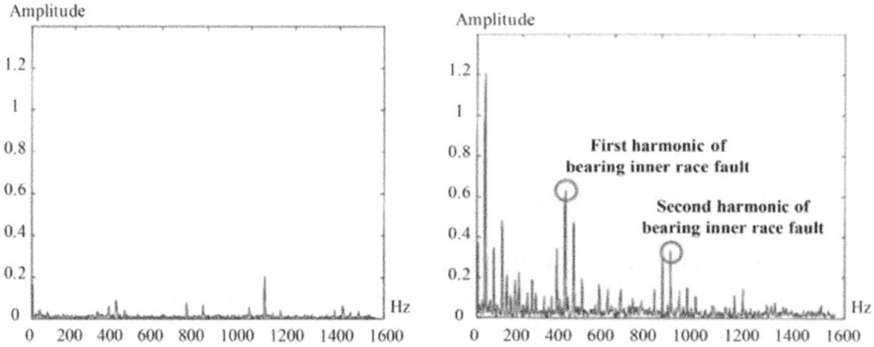

a. Without fault   *b. Inner race fault*

Figure 1.13. *Spectrum of the vibration signal envelope in the case of an inner race fault*

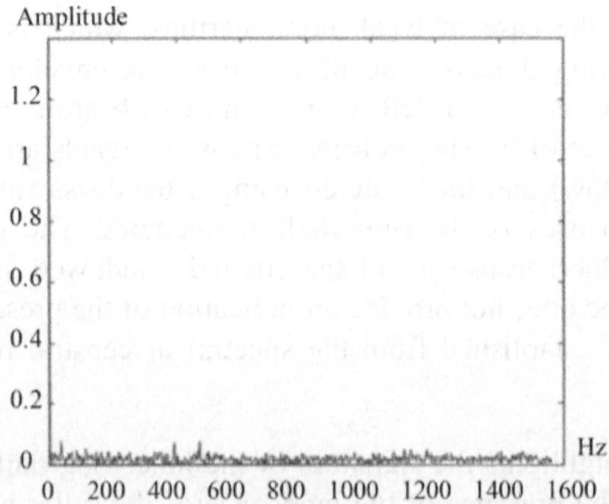

Figure 1.14. *Spectrum of the vibration signal envelope in the case of gear fault*

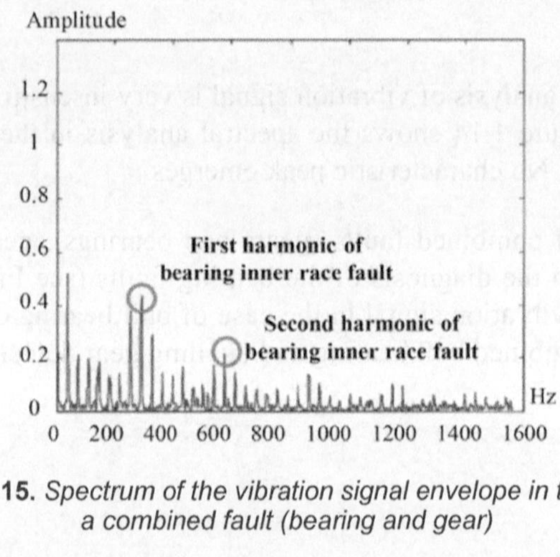

Figure 1.15. *Spectrum of the vibration signal envelope in the case of a combined fault (bearing and gear)*

1.5.2. *Frequency analysis*

Frequency analysis is an effective tool for detecting defects in rotating machinery [DAL 75]. This method used the principle of transforming time signals into their frequency equivalents. It has been found that the spectral content of the measured signals is sometimes more useful than the time domain for determining the condition of the gears, since the complex time signal can be decomposed into many frequency components. Thus, it is easy to focus on those frequencies that are useful for monitoring.

The fast Fourier transform (FFT) is the most common analysis technique used to make the transition to the spectral domain. However, this method is valid only when the spectral content of the measured signal does not change over time and so there are no variations in the speed of rotation of the machine. For machines operating at a constant speed, the frequencies of the excitations produced by each element of the system can be determined in a given frequency band, as shown in Figure 1.16. Thus, the sidebands generated by the amplitude or frequency modulations can provide useful information for detecting defects. Amplitude modulations are usually associated with natural gear tooth degradation that produces pulsations at a rate equal to the speed of rotation. On the other hand, frequency modulations are attributed to manufacturing defects and sometimes to cracking defects [TC 99].

However, the spacing of the sidebands depends on the periodic properties of the load and the drive shaft. It is sometimes difficult to extract useful features directly from the signal spectrum using FFT. When the signal-to-noise ratio is low and the signal spectrum has a large number of frequency components because of the complexity of the system, it becomes very difficult to distinguish the frequency peaks relative to the defects from other sources. This is an important limitation of this method for fault detection [DID 04].

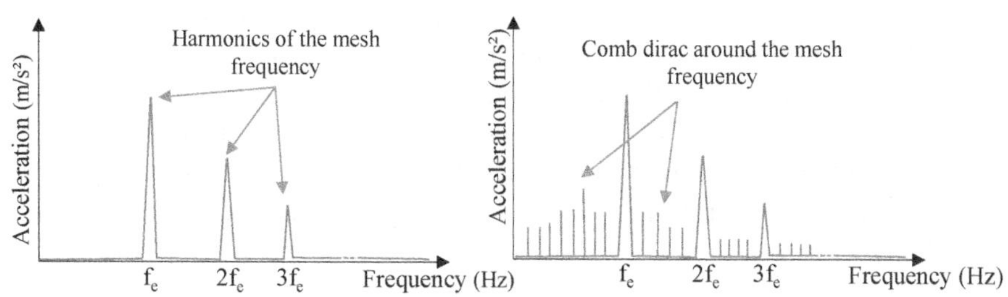

Figure 1.16. *Spectrum of the vibration signal of gears in (a) healthy and (b) faulty conditions*

Among the most widely used frequency analysis techniques, we cite the demodulation spectrum of the envelope amplitude [FL 14]. It is analogous to temporal envelope analysis and is divided into two techniques. The first is based on the filtering, the square of the signal followed by its FFT. The second type represents the evaluation of existing modulations based on the analytic function, which is determined from the absolute value of the Hilbert transform H(x)(f) in the frequency domain, as shown by the following equation:

$$H(x)(f) = -j\,\mathrm{sgn}(f)X(f). \tag{1.3}$$

With sgn(f) being the sign function in frequency and X(f) the Fourier transform of the signal x(t).

Also, cepstral analysis is a technique that proves effective in the detection of gear defects [OS 04]. The cepstrum is defined as the inverse of the spectrum and focuses on amplitudes, as given by

$$C(\tau) = -F^{-1}[\log 10 S(f)]. \tag{1.4}$$

Thus, this method facilitates the separation of different periodicities of the sideband groups linked to the gears. It is practiced in various applications such as signals measured with angular sampling from optical encoders. It reduces the effect of speed variations for a cepstrum under temporal sampling [BGN 97].

1.5.3. *Time–frequency analysis*

Time domain and frequency domain signal analyses produce signal characteristics relating exclusively to their respective domains. When the signal in the time domain is transformed to the frequency domain, the detailed information concerning the time domain is lost. Thus, it is often difficult to clearly detect the symptoms of gear defects if the synchronous average, for example, is used in isolation. This kind of technique is also sometimes unable to detect and differentiate between defects, especially multiple defects presenting simultaneously in the gear mechanism [LS 01].

Many researchers are interested in signal analysis in the time–frequency domain in order to have a more complete representation of the signal. The major differences between these transforms are their respective temporal and frequency resolutions. Thus, time–frequency analysis covers both time and frequency domains in order to reduce the compromise between their resolutions. Nonstationary and modulated signals are most appropriate for the time–frequency distribution, the principle of which is to give the distribution of the signal energy in a two-dimensional time–frequency space [BS 95]. Thus, the study of nonstationary signals is of fundamental interest for time–frequency analysis.

One of the most commonly used methods in time–frequency analysis is the short-term Fourier transform (STFT). This method tracks the change in the frequency content as a function of time, which implies that the fault becomes time definable. This transform is based on a sliding window that performs the windowing of the signal around a particular time and calculates the Fourier transform at each moment of the signal. STFT requires significant computing time to obtain a higher quality resolution [LEE 13]. In addition, this technique depends substantially on the width of the windowing used, which greatly limits its scope. Thus, despite its efficiency, in general, STFT is not optimal in the context of nonstationary signals.

In order to overcome the problems of STFT, wavelet analysis is proved to be an effective method for gear-time monitoring in the time–frequency domain. Unlike STFT which uses constant time windows, wavelet analysis uses narrow high-frequency time windows and wide low-frequency time windows [AFGL 96]. The wavelet transform depends on the function of time and scale, but it also uses a self-adaptive window based on a wavelet function of which the duration varies inversely proportional to the scale factor. Thus, it is a robust method for the analysis of transient signals. Indeed, the transient components generated by early gear defects are detectable from discrete and continuous transforms [GY 03]. However, it has been shown that although discrete wavelet analysis provides a nonredundant signal representation, the timescale mapping that it provides is limited and therefore relatively poor in information [HW 96].

1.6. Diagnostic methods

The diagnosis is defined as the process of characterization and localization of the defect after detection. This is an important phase in conditional preventive maintenance, as it decides the action to be taken based on the defect identified. The diagnosis takes place after the detection of the defect and aims at isolating it in order to locate it and to characterize it in terms of causes and effects. When the defect is still in its infant stage and when the system is still in a controllable state of operation, the diagnosis avoids untimely shutdowns and thus minimizes the resulting decrease in productivity. Therefore, the diagnosis of defects is a major research topic attracting the interest of industry and researchers [YAN 04, HS 17].

Many diagnostic techniques are reported in the literature. The study conducted by Venkatasubramanian *et al.* [VRYK 03a, VRYK 03b,

VRYK 03c] classifies these methods into two broad categories based on the prior knowledge process: (i) model-based methods approach and (ii) data-driven methods approach. The proposed classification does not take into account hybrid methods, which makes this proposal no exhaustive. The hierarchy of diagnostic techniques is shown in Figure 1.17.

The type of knowledge process used *a priori* is a determining factor for the classification of diagnostic methods. The prior knowledge required for diagnosis is the set of defects as well as the relationship between observations and failures. This knowledge is extracted from the representation of the system's process by a mathematical model. In this case, the approach is considered to be based on a model. On the other hand, it establishes the knowledge base directly from the past experience of the process in question. This approach is, therefore, centered around historical data that can be treated as knowledge *a priori* to the diagnosis, thus optimizing the process. This is called the data-driven diagnostic method [MIL 87].

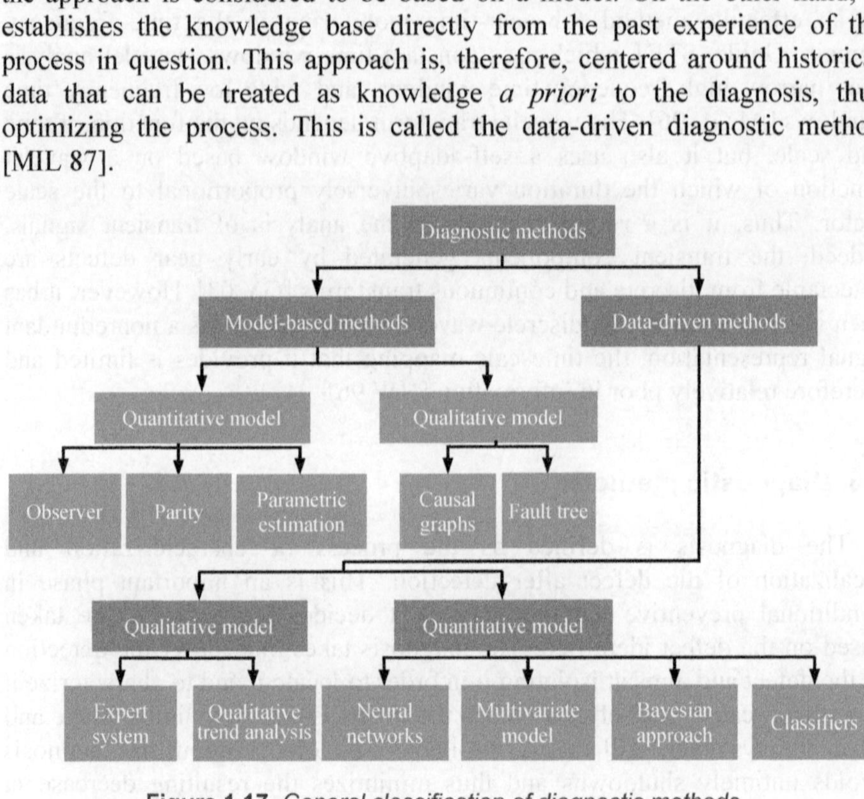

Figure 1.17. *General classification of diagnostic methods*

Methods based on the model approach classify *a priori* knowledge between qualitative and quantitative approaches. In the quantitative method, the understanding of the physics of the process is expressed in terms of mathematical relationships between the inputs and outputs of the system. However, in qualitative models, the relationship characterizing the physics of the process is represented as qualitative functions.

For a data-based approach, only the availability of a historical database is considered. There are different ways in which these data can be transformed and exploited in the diagnosis as prior knowledge. This process is known as the extraction of characteristics from historical data and is used to facilitate subsequent diagnosis. This process can be done either quantitatively or qualitatively. In the quantitative approach, this process is based on mathematical relationships that can be statistical or no statistical. On the other hand, the process of extracting characteristics qualitatively is based on qualitative functions.

1.6.1. *Model approach*

When historical data is unavailable and when it is possible to represent the system using a model, the model approach becomes the most appropriate.

1.6.1.1. *Quantitative methods*

Most quantitative approach models are based on residues. These methods need to generate irregularities between the actual behavior of the system and the estimation of the model. These irregularities, called residues, reflect the presence of a failure in the system and involve the selection of a decision rule for diagnosis. Quantitative models include the observer, the parity space and the parametric estimate.

1.6.1.1.1. Observer

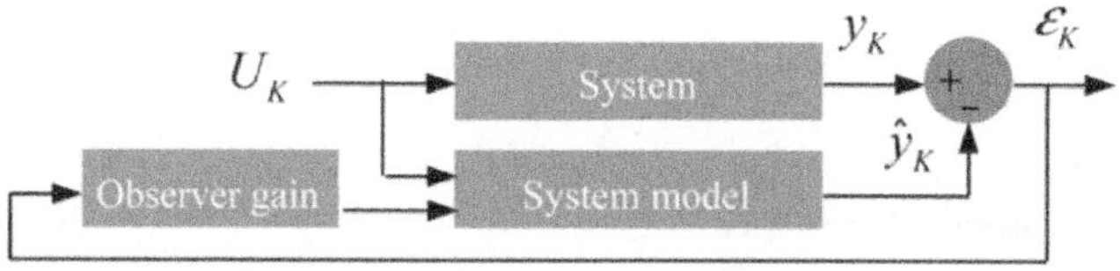

Figure 1.18. *Principle of the observer and the generation of the residue*

Observer-based methods consist of constructing a set of partial observations of the system in the form of inputs/outputs and state variables. The residue represents the difference between the outputs measured in the system and the values estimated by the model. When the residue value is nonzero, it indicates a divergence that reflects the appearance of a defect. This process can be performed using observers in a deterministic configuration [PC 97] or statistical filters like the Kalman filter in a stochastic configuration [SMB 00]. Figure 1.18 presents the principle of residue generation in the observer.

1.6.1.1.2. Parity space

Parity space transforms the system's state-space model to generate redundancy relationships between inputs and outputs that are measured variables. Analytical redundancy relationships are obtained by eliminating unobservable state variables from the model and are subsequently used to construct the fault indicators. This technique rearranges the structure of the model so that the best insulation of the defect can be obtained, which can be proved by checking the parity of the model with the measurements. The optimal isolation of the fault requires the generation of orthogonal residual vectors with respect to each other for different defects.

The parity space method was first developed by Potter and Sunman for static systems [PS 77] and was later adapted by Chow and Willsky to dynamic systems [CW 84]. Parity relationships and observers have been shown to be equivalent to the level of the generated residue [GER 91]. Thus, both methods have similar problems, including noise management and system errors [DGT 99]. Figure 1.19 schematizes the principle of the generation of the residue of the parity space.

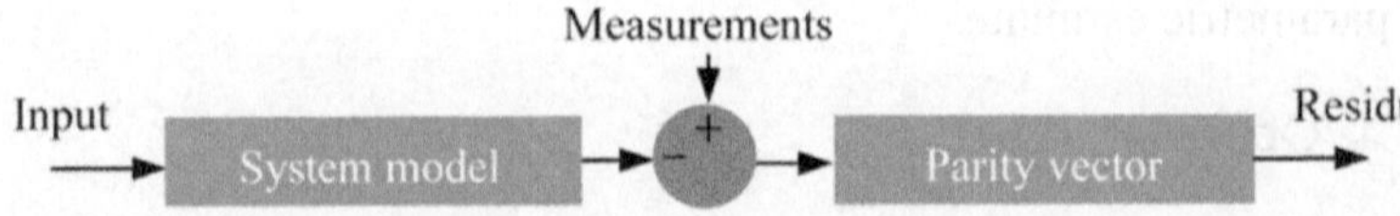

Figure 1.19. *Principle of the parity space and the generation of the residue*

1.6.1.1.3. Parametric estimation

This method is based on the principle that faults influence the physical parameters of the system. The comparison between the estimated parameters and the actual parameters indicates the appearance of the fault. Parametric estimation has the advantage of quantifying the deviation, which is useful for further analysis of the defect. However, one of the major drawbacks of this method is the need for a constantly excited physical system. This method has been proposed for the detection of defects in different system configurations [DWCX 16, WAN 16, XCX 15]. Figure 1.20 represents the principle of parametric estimation.

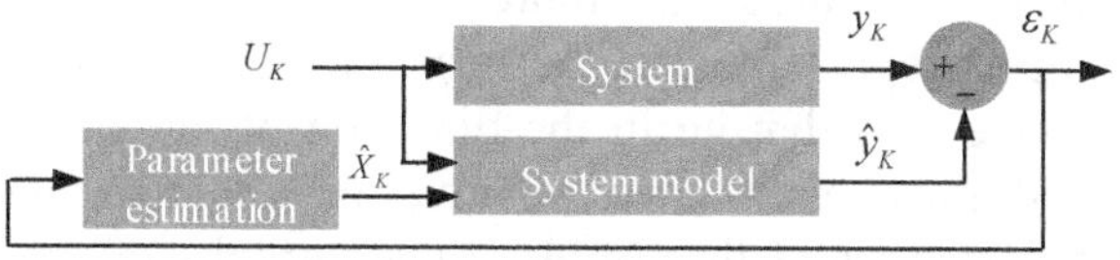

Figure 1.20. *Representation of the principle of parametric estimation*

1.6.1.2. *Qualitative methods*

The qualitative methods of the model approach are based on different forms of qualitative knowledge used in diagnosis. These methods are classified into fault tree, causal graph and qualitative physics.

1.6.1.2.1. Fault tree

The fault tree synthesizes all combinations of events that lead to the fault. It is a representation that reflects the behavior of the defect in the form of a visual diagram. The diagram is composed of a set of rules and logical symbols, allowing analysis of complex systems. The high event, representing the top of the tree, is the defect that needs to be analyzed. On the other hand, the low event is connected to the high event via intermediate events, called nodes, which represent the progress of the fault in the component being studied. Events of a given level are connected to the following levels through logical gates (AND, OR, etc.). Fault trees have been used in several reliability analysis studies [KL 86, LP 77]. Figure 1.21 shows the principle of the fault tree.

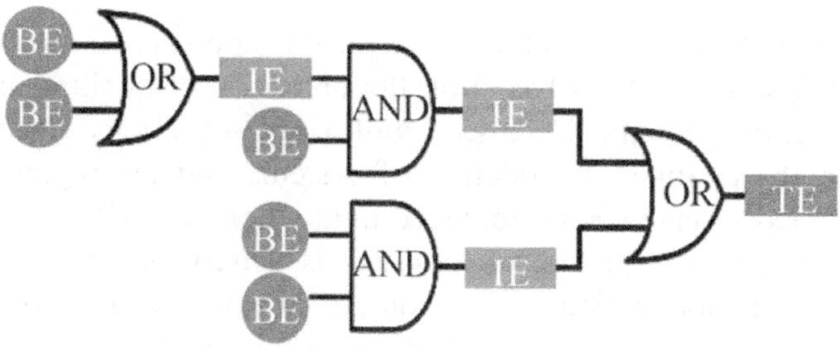

Figure 1.21. *Basic structure of a fault tree*

1.6.1.2.2. Causal graph

The causal model is a graph with orientation that allows connecting the defects to their symptoms based on cause and effect. The diagnosis consists of locating all possible sources of failure from the sensor data. The cause–effect relationship is represented in the form of digraphs with signs. These are graphs that have arcs that go in the direction of the cause nodes to the effect nodes, and these arcs have positive or negative signs based on comparison with established thresholds. These graphs effectively describe qualitative models. They are interesting because they require little information to establish the network of graphs and thus make the diagnosis from the causal knowledge of the system [YSX 12, LWX 16]. However, although the causal model is easy to implement, it is possible that there is a considerable loss of information as well as the ability to distinguish between defects when it is a complex system [BBLM 14]. Figure 1.22 schematizes the principle of digraphs with signs.

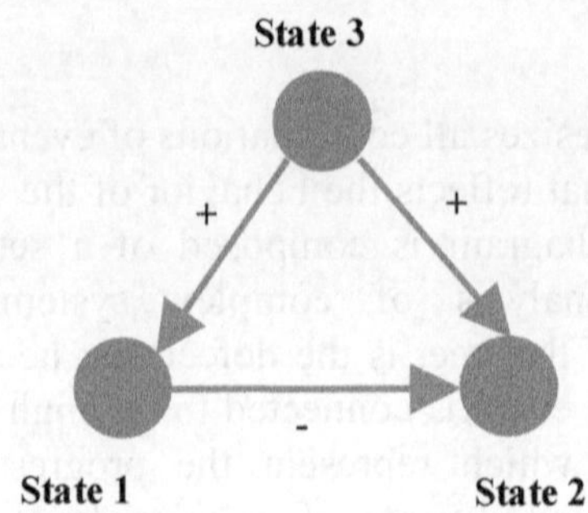

Figure 1.22. *Basic principle of digraphs with signs*

1.6.1.2.3. Qualitative physics

The knowledgebase of qualitative physics is approached in two different ways. The first approach is based on the obtention of qualitative equations from differential equations called confluence equations. Many studies focused on the qualitative modeling of systems and the representation of causal knowledge have been referenced in the literature [KB 84, IS 86]. The second approach of qualitative physics is based on the obtention of qualitative behavior acting as a source of knowledge from ordinary differential equations. These are equations where partial derivatives do not intervene, unlike partial differential equations. In terms of applications of qualitative models, qualitative simulation is used to predict the qualitative properties of the system [KK 93]. Also, qualitative process theory is widely used in the prediction of qualitative behavior [GU 91].

1.6.2. *Data-driven approach*

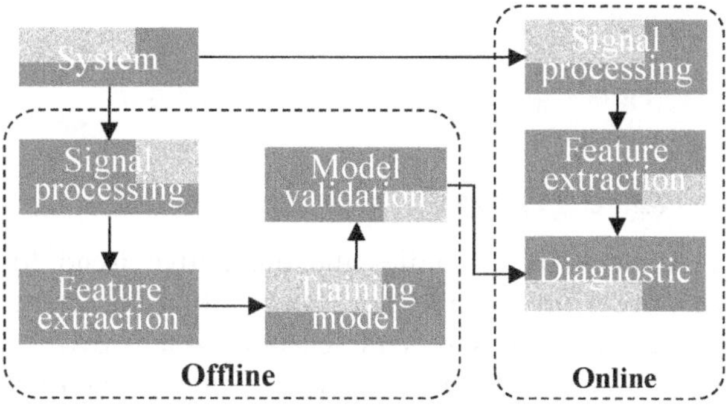

Figure 1.23. *Basic architecture of the diagnosis in the data-driven approach*

Unlike the model approach where prior knowledge is required, in the data-based approach, only the analysis of recorded data is required. This approach establishes diagnostic processes that do not rely on the physical modeling of the system studied. Methods of the data-based approach can be either quantitative or qualitative. Figure 1.23 shows the diagnostic architecture based on this approach.

1.6.2.1. *Qualitative methods*

The expert system and qualitative analysis are the two methods used for the extraction of qualitative historical information.

1.6.2.1.1. Expert system

The expert system is an intelligent computer program using a knowledgebase and an inference engine to solve complex problems. It is usually a specialized system using a knowledgebase combined with an "if–then" rule indicating how to apply that knowledge. Many researchers have applied expert systems for diagnostic problems [GU 91, BL 93, TS 97]. The advantage of this system is its simplicity of implementation. Nevertheless, it is unable to detect new defects without knowing their signatures. Thus, this system needs to be continually updated as the system configuration evolves [RIC 87]. Figure 1.24 shows the basic architecture of an expert system.

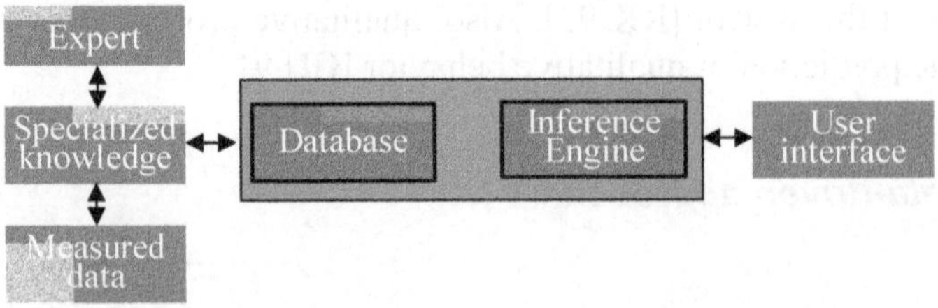

Figure 1.24. *Architecture of an expert system*

1.6.2.1.2. Qualitative analysis

Qualitative trend analysis studies the qualitative trend to explain the different events occurring within the system. It is based on information extracted from trends in measurement sensors through two steps. The first step is the identification of trends in the measurements, and the second step is the interpretation of trends in terms of defect scenario. Thus, these methods allow the diagnosis as well as the prediction of future states of the system [VRYK 03c, GCD 15].

1.6.2.2. *Quantitative methods*

Methods centered around the extraction of quantitative information from the historical database are broadly divided into statistical and nonstatistical methods. Principal component analysis (PCA) and partial least-squares (PLS) as well as classification of statistical trends form the kernel of statistical methods for extracting diagnostic features. On the other hand, nonstatistical methods are those based on supervised learning, especially neural networks.

1.6.2.2.1. Statistical methods

Multidimensional statistical methods such as the PCA and PLS have been applied to the diagnosis and detection of defects. The PCA method is a technique based on features reduction. The principle of this method is equivalent to minimizing the mean-squared error between measured data and their projections in the reduced feature space. The initial features are transformed into an uncorrelated set of features. These represent the variations in the system and can thus be used for fault detection [MK 95, WG 96].

Another multidimensional statistical method is the PLS. This method aims to reduce the size of the feature space by making the projection that best describes the measured data in terms of least-squares in order to better preserve the discriminant information. The purpose of this representation is to find a linear regression model by projecting the measured data in a new feature space [QIN 98, QM 92].

The classification is also a part of statistical methods used for diagnosis. The purpose of the classification is to categorize measured data, called observations, into classes representing the health states of the system (see Figure 1.25). An observation is represented by a set of features. These features allow for the construction of a representation space grouping similar observations. Similar observations are grouped in the same class. It is, therefore, possible to diagnose the state of the system by identifying the membership class of the acquired data.

There are two types of classification: unsupervised and supervised. The choice of a classification type depends on the availability of the training data. If we have a database of observations offering *a priori* knowledge of the corresponding membership class, we speak of supervised classification; otherwise, it is unsupervised classification.

Several unsupervised classification algorithms have been developed. One example is the K-means algorithm. This algorithm consists of defining centroids representing the gravity centers of the classes and then constructing groups of observations around the appropriate centers. We can also cite the fuzzy c-means algorithm, which defines a degree of membership indicating the similarity degree of grouped observations corresponding to the decision rule [PS 15].

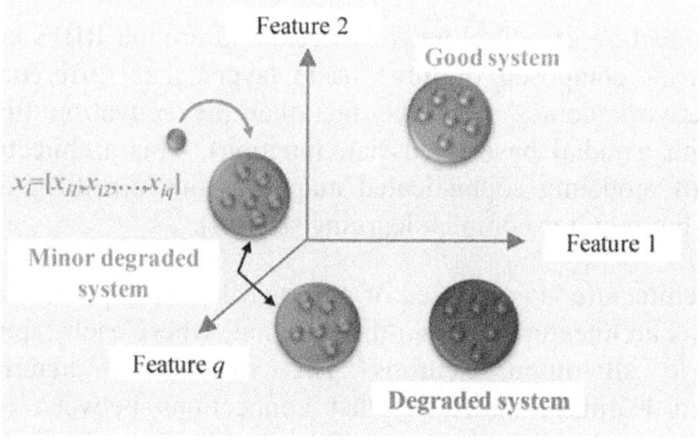

Figure 1.25. *Principle of the classification method*

For the supervised classification, among the applied algorithms, we can cite the support vector machine. This model aims to establish the boundary separating the different classes so that the margin is at its maximum. The observations used to determine the distance of the margin are called support vectors [YY 12].

1.6.2.2.2. Learning-based methods

Neural networks are the most frequently used learning-based methods in diagnosis. The basic principle of the neural network is the evaluation of the similarity between the inputs and a training set, followed by a grouping based on this evaluation. Although neural networks are part of supervised classification methods as presented before, their learning process is particularly exploited for diagnosis. Indeed, the neural networks used in fault monitoring can be classified into two categories: (i) the network architecture and (ii) the learning strategy [MB 15].

From an architectural point of view, neural networks can have several forms used in diagnostics such as multilayer perceptrons (MLPs), radial-based functions (RBFs) or self-organization Kohonen map (SOM). The basic principle of a single neuron is the combination of inputs to a set of weights through a sum function, as well as an activation function subject to a certain threshold to obtain the output. When the activation function is a sign function, we speak of a simple perceptron. Then by assigning a set of simple perceptrons to successive layers, including an input layer, an output layer and one or more intermediate hidden layers, an MLP is obtained [PAO 89].

The architecture of neural networks centered around RBFs consists of a neural network composed of three basic layers. The difference with the classical network comes from the fact that the activation function is a function with a radial base (Gaussian function). This architecture has the advantage of modeling sophisticated mappings that are difficult to obtain with MLPs but requires complex learning [BL 88].

SOM architecture is composed of only two layers: input and output. The layers of this architecture are two-dimensional, where each input neuron is connected to all output neurons. This structure is mainly used in unsupervised learning, indicating that connections between neurons are unknown. Thus, it has the advantage of being able to automatically build the representation of the data for diagnosis [UFKN 13].

1.7. Conclusion

In this chapter, a state of the art of gear fault diagnosis methods in the context of condition-based maintenance was presented. First, from the faults repartition occurring in the gearbox, it has been found that the gear defect is the predominant fault, thus justifying the interest of diagnosing it. Second, different types of damage appearing in gears, between defects distributed on all the teeth (wear and pitting) and localized defects on a tooth (flaking, cracking and seizing), were studied in order to characterize them for the diagnosis.

Then a study of the diagnosis methodology applied in the field was proposed. Hence, it has been shown that the general concept of diagnosis comprises of three main phases. First, the analysis which is the set of technical processes that ensures the extraction of health indicators (features). Then, the monitoring that exploits the fault indicators for detection. Finally, the diagnosis which is the identification of the detected defect for the characterization and orientation of the decision-making.

In the second section, the methods of analysis were presented: temperature analysis (temperature sensor and infrared thermography), oil analysis, acoustic analysis (acoustic emission) and vibration analysis. Other emerging methods, such as analysis of electromagnetic torque, the flux analysis and the analysis of electric currents, were also presented.

After that, the most frequently used monitoring methods in the literature were presented. The notion of "fault signature" has been introduced and it has been shown that these methods are classified into three major axes. First, the temporal analysis that groups together techniques such as statistical indicators, synchronous averaging and envelope analysis. Second, the frequency analysis that admits the Fourier transform, the demodulation of the amplitude spectrum of the envelope and cepstral analysis. Finally, time–frequency analysis that includes STFT.

Finally, it was a question of detailing the diagnostic methods which are divided into several classes between quantitative methods (observer, parity space and parametric estimation) and qualitative methods (fault tree, causal graph and qualitative physics) in the model approach, as well as quantitative (learning and statistical) and qualitative methods (expert system and qualitative analysis) in the data-driven approach.

1.8. References

[AFGL 96] F. Auger, P. Flandrin, P. Goncalves, and O. Lemoine. *Time–Frequency Toolbox*, Paris, France/Houston, TX: CNRS/RICE Univ, 1996.

[ARQ 09] P. Arques. Diagnostic *prédictif* et défaillances. Technical report, TECHNIP, 2009.

[AUG 02] D. Augeix. Analyse vibratoire des machines tournantes. *Techniques de l'Ingénieur, traité Génie Mécanique*, 5(145): 1–22, 2002.

[AYA 15] M. Ayad. Recherche de signatures robustes des défauts dans une machine tournante à travers l'analyse des signaux vibratoires. PhD thesis, Ferhat Abbas University, Setif, Algeria, 2015.

[BAD 99] M. El Badaoui. Contribution au diagnostic vibratoire des réducteurs complexes à engrenages par l'analyse cepstrale. PhD thesis, Jean Monnet University, Saint-Etienne, France, 1999.

[BGN 97] M. El Badaoui, F. Guillet, N. Nejjar, P. Martini, and J. Danière. Diagnostic d'un train d'engrenages par analyse cepstrale synchrone. In *seizième colloque GRETSI*, Grenoble, France, pp. 761–764, 1997.

[BL 93] W. Becraft and P. Lee. An integrated neural network/expert system approach for fault diagnosis. *Computers & Chemical Engineering*, 17:1001–1014, 1993.

[BBLM 14] B. Bouamama, G. Biswas, R. Loureiro, and R. Merzouki. Graphical methods for diagnosis of dynamic systems: review. *Annual Reviews in Control*, 38:199–219, 2014.

[BRE 02] C. Breneur. Eléments de maintenance préventive des machines tournantes dans le cas de défauts combinés d'engrenages et de roulements, INSA de Lyon, France, Lyon, PhD thesis, 2002.

[BL 88] D. Broomhead and D. Lowe. Radial basis functions, multi-variable functional interpolation and adaptive networks. In *Defense Technical Information Center Archive*, Royal signals and radar establishment Malvern, United Kingdom, pp. 1–223, 1988.

[BS 95] L. Burgess and T. Shimbel. What is the prognosis on your maintenance program. *Engineering & Mining Journal*, 2:1–14, 1995.

[CAS 03] R. Casimir. Diagnostic des défauts des machines asynchrones par reconnaissance des formes. PhD thesis, Ecole centrale de Lyon, France, 2003.

[CLW 12] K. Chen, X. Li, F. Wang, T. Wang, and C. Wu. Bearing fault diagnosis using wavelet analysis. In *International Conference on Quality, Reliability, Risk, Maintenance, and Safety Engineering*, Chengdu, 2012, pp. 699–702.

[CW 84] E.Y. Chow and A.S. Willsky. Analytical redundancy and the design of robust failure detection systems. *IEEE Transactions on Automatic Control*, 29:603–614, 1984.

[DAL 75] G. Dalpiaz. Gear fault monitoring comparison of vibration analysis techniques. *Italian National Research Council in Prentice Hall International*, 27:623–632, 1975.

[DID 04] G. Didier. Modélisation et diagnostic de la machine asynchrone en présence de défaillances. PhD thesis, Henri Poincaré University, Nancy, France, 2004.

[DGT 99] X. Ding, L. Guo, and T. Jeinsch. A characterization of parity space and its application to robust fault detection. *IEEE Transactions on Automatic Control*, 44:337–343, 1999.

[DWCX 16] F. Ding, X. Wang, Q. Chen, and Y. Xiao. Recursive least squares parameter estimation for a class of output nonlinear systems based on the model decomposition. *Circuits, Systems, and Signal Processing*, 35:3323–3338, 2016.

[DBLR 04] J.P. Dron, F. Bolaers, and L. Rasolofondraibe. Improvement of the sensitivity of the scalar indicators (crest factor, kurtosis) using a de-noising method by spectral subtraction: application to the detection of defects in ball bearings. *Journal of Sound and Vibration*, 270(1):61–73, 2004.

[FL 14] Z. Feng and M. Liang. Fault diagnosis of wind turbine planetary gearbox under nonstationary conditions via adaptive optimal kernel time–frequency analysis. *Renewable Energy*, 66:468–477, 2014.

[FWZ +16] K. Feng, K. Wang, M. Zhang, Q. Ni, and M. Zuo. A diagnostic signal selection scheme for planetary gearbox vibration monitoring under non-stationary operational conditions. *Measurement Science and Technology*, 28:12, 2016.

[GCD 15] Z. Gao, C. Cecati, and S. Ding. A survey of fault diagnosis and fault tolerant techniques. Part I: Fault diagnosis with model-based and signal-based approaches. *IEEE Transactions on Industrial Electronics*, 62(6):3757–3767, 2015.

[GER 91] J. Gertler. Analytical Redundancy Methods in Fault Detection and Isolation - Survey and Synthesis, In *IFAC Proceedings Volumes*, 24(6):9–21, 1991.

[GY 03] V. Giurgiutiu and L. Yu. Comparison of short-time Fourier transform and wavelet transform of transient and tone burst wave propagation signals for structural health monitoring. *International Workshop on Structural Health Monitoring, Stanford University*, Stanford, pp.1267–1274, 2003.

[GU 91] S. Grantham and L. Ungar. Comparative analysis of qualitative models when the model changes. *AICHE Journal*, 37:931–943, 1991.

[HW 96] J. K. Hammond and P. R. White. The analysis of non-stationary signals using time–frequency methods. *Journal of Sound and Vibration*, 190:419–447, 1996.

[HS 17] Y. Hawwari and A. Soualhi. A survey about diagnosis and prognosis methods used in prognostic and health management. Master's thesis, Jean Monnet University, Saint-Etienne, France, 2017.

[HOG 16] F. Hemmati, W. Orfali, and M.S. Gadala. Roller bearing acoustic signature extraction by wavelet packet transform, applications in fault detection and size estimation. *Applied Acoustics*, 104:101–118, 2016.

[HLC 00] T. Hu, B.C. Lu, and G.J. Chen. A rotary machinery fault diagnosis approach based on rough set theory. In *World Congress on Intelligent Control and Automation*, Hefei, China, pp. 589–685, 2000.

[HTS 15] Z. Hurtado, C. Tello, and J. Sarduy. A review on location, detection and fault diagnosis in induction machines. *Journal of Engineering Science and Technology*, 8:185–195, 2015.

[IS 86] Y. Iwasaki and H. Simon. Causality in device behavior. *Journal of Artificial Intelligence*, 29:3–32, 1986.

[KK 93] H. Kay and B. Kuipers. Numerical behavior envelopes for qualitative models. *Proceedings of AAAI*, 93:606–613, 1993.

[KL 86] B.E. Kelly and F.P. Lees. The propagation of faults in process plants: 2. Fault tree synthesis. *Reliability Engineering*, 16:39–62, 1986.

[KHC 15] S. H. Kia, H. Henao, and G.A. Capolino. Trends in gear fault detection using electrical signature analysis in induction machine-based systems. In *IEEE Workshop on Electrical Machines Diagnosis*, Torino, Italy, pp. 297–303, 2015.

[KB 84] J. Kleer and S. Brown. A qualitative physics based on confluences. *Journal of Artificial Intelligence*, 24:7–83, 1984.

[LP 77] S. Lapp and G. Powers. Computer-aided synthesis of fault-trees. *IEEE Transactions on Reliability*, 26:2–13, 1977.

[LEE 13] J. Lee. Sound and vibration signal analysis using improved short-time Fourier representation. *International Journal of Automotive and Mechanical Engineering*, 7:811–819, 2013.

[LEE 08] V. Leemans. Les méthodes et outils de diagnostic et de controle appliqués pour la maintenance et la maitrise des risques – Application de la thermographie infrarouge au diagnostic. Technical report, ENIM, 2008.

[LS 01] S. Liu and W. Shi. Rough set based intelligence diagnostic system for valves in reciprocating pumps. *IEEE International Conference on Systems, Man and Cybernetics*, 1:353–358, 2001.

[LWX 16] Y. Liu, G. Wu, C. Xie, Z. Duan, M. Peng, and M. Li. A fault diagnosis method based on signed directed graph and matrix for nuclear power plants. *Nuclear Engineering and Design*, 297:166–174, 2016.

[LOU 04] S.J. Loutridis. Damage detection in gear systems using empirical mode decomposition. *Engineering Structures*, 26:1833–1841, 2004.

[MK 95] J. MacGregor and T. Kourti. Statistical process control of multivariate processes. *Control Engineering Practice*, 3:403–414, 1995.

[MS 85] P.D. McFadden and J.D. Smith. Model for the vibration produced by multiple point defects in a rolling element bearing. *Journal of Sound and Vibration*, 98:263–273, 1985.

[MIG 11] C. Migeon. Emission acoustique et analyse vibratoire pour l'étude des défauts de roulements pour différents régimes moteurs, Conservatoire national des arts et métiers, PhD thesis, Champagne-Ardenne, France, 2011.

[MIL 87] R. Milne. Strategies for diagnosis. *IEEE Transactions on Systems, Man and Cybernetics*, 17:333–339, 1987.

[MB 15] M. Msaaf and F. Belmajdoub. Application des réseaux de neurone de type feedforward dans le diagnostic statique. In *Conférence Internationale sur la Conception et Production Integré* Tanger, Maroc, pp. 1–6, 2015.

[NY 10] G. Niu and B. Yang. Intelligent condition monitoring and prognostics system based on data-fusion strategy. *Expert Systems with Applications*, 37:8831–8840, 2010.

[OS 04] Al.V. Oppenheim and R.W. Schafer. From frequency to quefrency: A history of the cepstrum. *IEEE Signal Processing Magazine*, 21:1–12, 2004.

[PAO 89] Y. Pao. *Adaptive Pattern Recognition and Neural Networks*. AddisonWesley Longman Publishing Co., Boston, 1989.

[PC 97] R.J. Patton and J. Chen. Observer-based fault detection and isolation: robustness and applications. *Control Engineering Practice*, 5:671–682, 1997.

[PS 15] E. Pazouki and S. Choi. Fault diagnosis and condition monitoring of bearing using multisensory approach based fuzzy-logic clustering. *IEEE International Conference on Electric Machines and Drives*, 2:1412–1418, 2015.

[PS 77] E. Potter and M. Sunman. Threshold-less redundancy management with arrays of skewed instruments. *AGARDOGRAPH*, 224:1–15, 1977.

[PCE 13] M.D. Prieto, G. Cirrincione, A.G. Espinosa, J.A. Ortega, and H. Henao. Bearing faults detection by a novel condition monitoring scheme based on statistical-time features and neural networks. *IEEE Transactions on Industrial Electronics*, 60:3398–3407, 2013.

[QIN 98] S. Qin. Recursive PLS algorithms for adaptive data modeling. *Computers & Chemical Engineering*, 22:503–514, 1998.

[QM 92] S. Qin and T. McAvoy. Nonlinear PLS modeling using neural networks. *Computers & Chemical Engineering*, 16:379–391, 1992.

[RIC 87] S. Rich. Model-based reasoning in diagnostic expert systems for chemical process plants. *Computers & Chemical Engineering*, 11:111–122, 1987.

[SMB 00] M.S.N. Said, M.E.H. Benbouzid, and A. Benchaib. Detection of broken bars in induction motors using an extended Kalman filter for rotor resistance sensorless estimation. *IEEE Transactions on Energy Conversion*, 15(1):66–70, 2000.

[SHKB 95] R.R. Schoen, T.G. Habetler, F. Kamran, and R.G. Bartheld. Motor bearing damage detection using stator current monitoring. *IEEE Transactions on Industry Applications*, 31:1274–1279, 1995.

[TC 99] N. Tandon and A. Choudhury. A review of vibration and acoustic measurement methods for the detection of defects in rolling element bearings. *Tribology International*, 32:469–480, 1999.

[TS 97] E. Tarifa and N. Scenna. Fault diagnosis, direct graphs, and fuzzy logic. *Computers & Chemical Engineering*, 21:649–654, 1997.

[THO 02] M. Thomas. *Fiabilité, maintenance prédictive et vibration des machines*. PhD thesis, Université du Québec, Ecole de Technologie Supérieure, 2002.

[TMMZ 12] D. Tobon-Mejia, K. Medjaher, and N. Zerhouni. CNC machine tool's wear diagnostic and prognostic by using dynamic Bayesian networks. *Mechanical Systems and Signal Processing*, 28:167–182, 2012.

[TCZ 15] K. L. Tsui, N. Chen, Q. Zhou, Y. Hai, and W. Wang. Prognostics and health management: a review on data driven approaches. *Hindawi Publishing Corporation Mathematical Problems in Engineering*, 15:1–17, 2015.

[UFKN 13] A. Urmos, M. Farkas, L. T. Koczy, and A. Nemcsics. Quantum structure classification by Kohonen self-organizing map and by fuzzy c-means algorithm. In *IEEE International Conference on System Science and Engineering*, Budapest, pp. 1–12, 2013.

[UA 82] S. Usui and I. Amidror. Digital low-pass differentiation for biological signal-processing. *IEEE Transactions on Biomedical Engineering*, 29:686–693, 1982.

[VPM 15] J. Vasselin, M. Petit, and J. Mocq. L'apport de l'analyse des courants d'alimentation des moteurs pour la maintenance conditionnelle des machines tournantes: Mise en oeuvre et retours d'experience. In *22ème Congrès Français de Mécanique*, Lyon, France, pp. 1–6, 2015.

[VRYK 03a] V. Venkatasubramanian, R. Rengaswamy, K. Yin, and S. N. Kavuri. A review of process fault detection and diagnosis. Part I: Quantitative model based methods. *Computers & Chemical Engineering*, 27:293–311, 2003.

[VRYK 03b] V. Venkatasubramanian, R. Rengaswamy, K. Yin, and S. N. Kavuri. A review of process fault detection and diagnosis. Part II: Qualitative models and search strategies. *Computers & Chemical Engineering*, 27:313–326, 2003.

[VRYK 03c] V. Venkatasubramanian, R. Rengaswamy, K. Yin, and S. N. Kavuri. A review of process fault detection and diagnosis. Part III: Process history based methods. *Computers & Chemical Engineering*, 27:327–346, 2003.

[WAN 16] D. Wang. Hierarchical parameter estimation for a class of MIMO Hammerstein systems based on the reframed models. *Applied Mathematics Letters*, 57:13–19, 2016.

[WG 96] B. Wise and N. Gallagher. The process chemometrics approach to process monitoring and fault detection. *Journal of Process Control*, 6:329–348, 1996.

[XCX 15] L. Xu, L. Chen, and W. Xiong. Parameter estimation and controller design for dynamic systems from the step responses based on the newton iteration. *Nonlinear Dynamics*, 79:2155–2163, 2015.

[YAN 04] Q. Yang. Model-based and data driven fault diagnosis methods with applications to process monitoring. PhD thesis, Electrical Engineering and Computer Sciences, Case Western Reserve University, 2004.

[YSX 12] F. Yang, S.L. Shah, and D. Xiao. Signed directed graph based modeling and its validation from process knowledge and process data. *International Journal of Applied Mathematics and Computer Science*, 22:41–53, 2012.

[YY 12] A. Younus and B. Yang. Intelligent fault diagnosis of rotating machinery using infrared thermal image. *Expert Systems with Applications*, 39:2082–2091, 2012.

2

Techniques for Predicting Defects in Bearings and Gears

2.1. Introduction

Several studies using the pattern recognition approach for the diagnosis of faults in industrial systems are available [BMMB 05, MOU 09, PSSR 18, RDNWV 10, SCZX 04]. This chapter demonstrates the suitability of the pattern recognition principle for the prediction of faults in bearings and gears. For this purpose, we start with defining the principle of pattern recognition. A pattern is represented by a class composed of observations. An observation, carried out on a bearing or gear, is characterized by a set of q-features, where q is the number of features used to follow the degradation of the system (bearings, gears). This observation is represented by a point in a q-dimensional space called representation space, as defined by different monitoring features. These features are extracted by analyzing signals measured by sensors placed on the system. Since the features are real numbers, an observation denoted x_i is defined by a feature vector $x_i = [x_{i1}, x_{i2}, \ldots, x_{iq}]$ (see Figure 2.1).

A group of similar observations is denoted a class (Ω_i), where i represents the ith class of the system. In the case of the prognosis, a class (Ω_i) illustrates a degradation state (DS) of the system. Ideally, a class grouping identical observations should be represented by a single point in the representation space. Nevertheless, disturbances (noise, sensor accuracy, etc.) can appear. This implies the apparition of approximately similar observations in the same class. The influence of the noise on a set of M-classes (Ω_1, $\Omega_2, \ldots, \Omega_M$)

Chapter written by Abdenour SOUALHI and Hubert RAZIK.

representing the M-degradation states of the system results in representing these classes with areas in the representation space, grouping similar observations (see Figure 2.2). The goal is to use this representation space to build prediction models based on hidden Markov models (HMMs) to predict the imminence of the next DS.

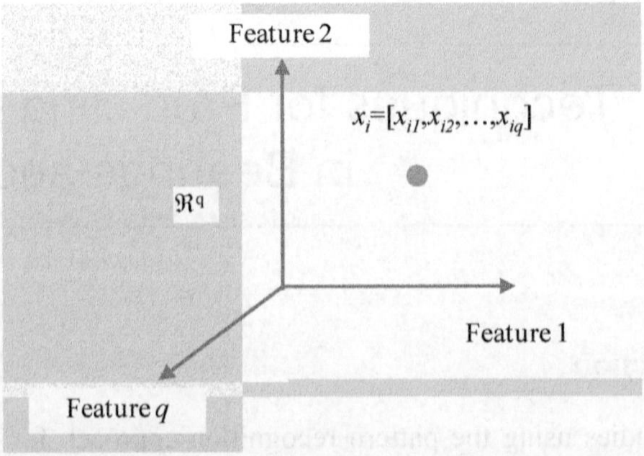

Figure 2.1. *Representation of an observation in a q-dimensional space. For a color version of the figures in this book, see www.iste.co.uk/soualhi/electrical1.zip*

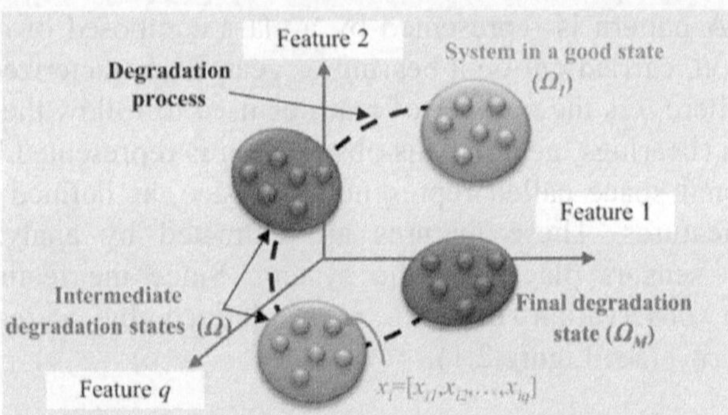

Figure 2.2. *Visualization of different DSs of a system by classes*

As shown in Figure 2.3, several steps are needed to build a prediction model. The signal processing step is presented in Chapter 1.

Feature vector construction: This step is used to build a set of features sensitive to the degradation of the system. The feature vector is initially composed by a set of q'-features obtained by applying different signal processing techniques (time, frequency, time–frequency analysis; see Chapter 1) on the measured signals collected from sensors placed on the system.

Feature selection: Determination of the representation space consists of extracting from an initial set of q'-features a subset of relevant q-features ($q<q'$) sensitive to the degradation of the system. The acquired observations contain a number of irrelevant features. Irrelevant features are not sensitive to the degradation. The determination of the representation space can be obtained by separability between classes and the compactness of each class.

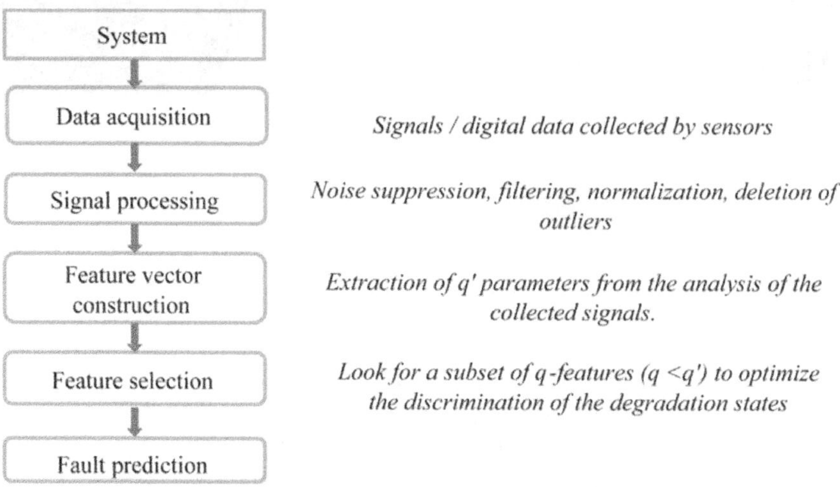

Figure 2.3. *Steps necessary for the prediction of bearing/gear faults by pattern recognition*

The feature vector construction step is used to build a set of features to show the effectiveness of using the pattern recognition approach to predict the degradation of bearings and gears. Consider two experimental cases.

2.1.1. *Bearing degradation*

Bearing degradation was carried out on a test bench designed by Rexnord Corp. The test bench is comprised of four roller bearings (ZA-2115) placed on a shaft, as shown in Figure 2.4. The characteristics of these bearings are detailed in Table 2.1.

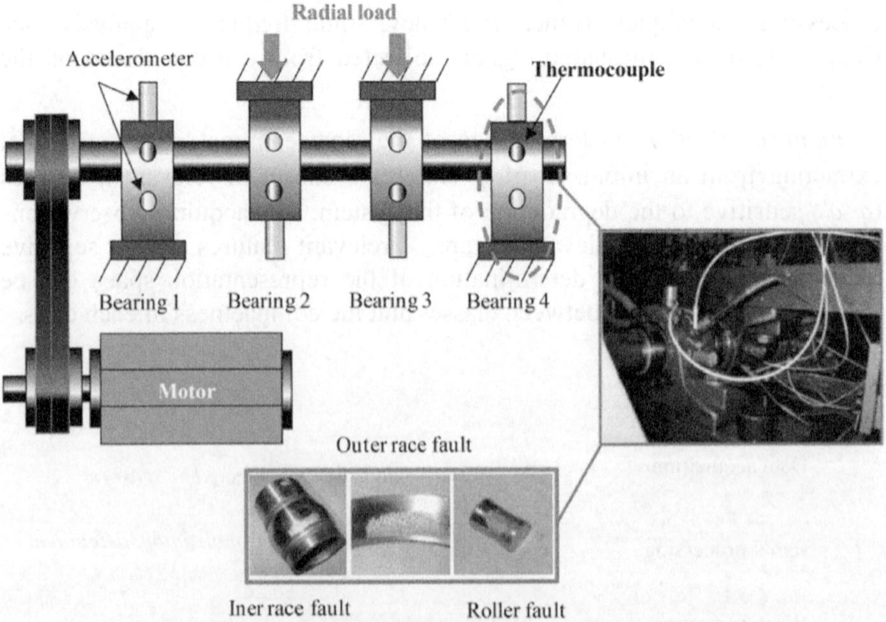

Figure 2.4. *Aging roller bearings test bench [QLLY 06]*

Bearing diameter (mm)	8.4
Roller number	16
Primitive diameter (mm)	71.62
Contact angle (°)	15.17

Table 2.1. *Bearing characteristics*

The rotation speed of the tested bearings is fixed at 2,000 rpm. The aging is accelerated by a spring mechanism generating a radial load of 26.6 kN on the tested bearings. An accelerometer was placed on the horizontal (X) and vertical (Y) axes of each bearing. The vibratory signals were recorded every 10 min with a sampling frequency of 20 kHz.

The signal processing techniques described in Chapter 1 are applied to extract the degradation indicators (features) from the vibration signals of bearings (1, 2, 3 and 4). In this case, we will use statistical features. These features will be used to identify the different DSs of bearings. These features are summarized in Table 2.2.

Energy	$E = \sum_{\ell=1}^{N} x_\ell^2$	γ_1
Peak amplitude	$S_{\text{peak}} = \sup_{1 \le \ell \le N} \lvert x_\ell \rvert$	γ_2
Mean value	$\overline{x} = \frac{1}{N}\sum_{\ell=1}^{N} x_\ell$	γ_3
Crest factor	$F_c = \frac{S_{\text{crête}}}{S_{\text{eff}}}$	γ_4
Power	$P = \frac{1}{N}\sum_{\ell=1}^{N} x_\ell^2$	γ_5
RMS value	$S_{\text{RMS}} = \sqrt{\frac{1}{N}\sum_{\ell=1}^{N}(x_\ell)^2}$	γ_6
Kurtosis	$S_{\text{kurt}} = \frac{(1/N)\sum_{\ell=1}^{N}(x_\ell - \overline{x})^4}{S_{\text{eff}}^4}$	γ_7
Factor K	$F_K = S_{\text{peak}} S_{\text{RMS}} = F_c S_{\text{RMS}}^2$	γ_8

Table 2.2. *Statistical features extracted from the vibration signal*

Here $x(t)$ is the acquired signal, N the number of samples composing the signal, f_{ech} the sampling frequency and $x_\ell = x(t = \ell/f_{\text{ech}})$, with ℓ being a nonzero positive integer less than or equal to N.

Figure 2.5 shows the evolution of the degradation of bearings using features extracted from the vibration signals acquired from accelerometers placed on the horizontal axis. Note that bearings 1, 2, 3 and 4 are tested during 35 days.

From Figure 2.5, we can eliminate some irrelevant features. We can observe that factor K is not a good indicator to follow the degradation of bearings. This is demonstrated for bearing 4, where factor K starts to decrease after 44,640 min of testing. The same remark is also applied for the kurtosis, crest factor and peak amplitude. A decrease in the value of these features is observed for bearings 3 and 4 after 44,640 and 47,520 min of testing. For the energy of the vibratory signal, it is considered as the image of power. So, it is sufficient to choose one of them to observe the degradation of bearings. Finally, three relevant features are chosen to construct the representation space. These features are: the mean value, power and RMS $[\gamma_3, \gamma_5, \gamma_6]$.

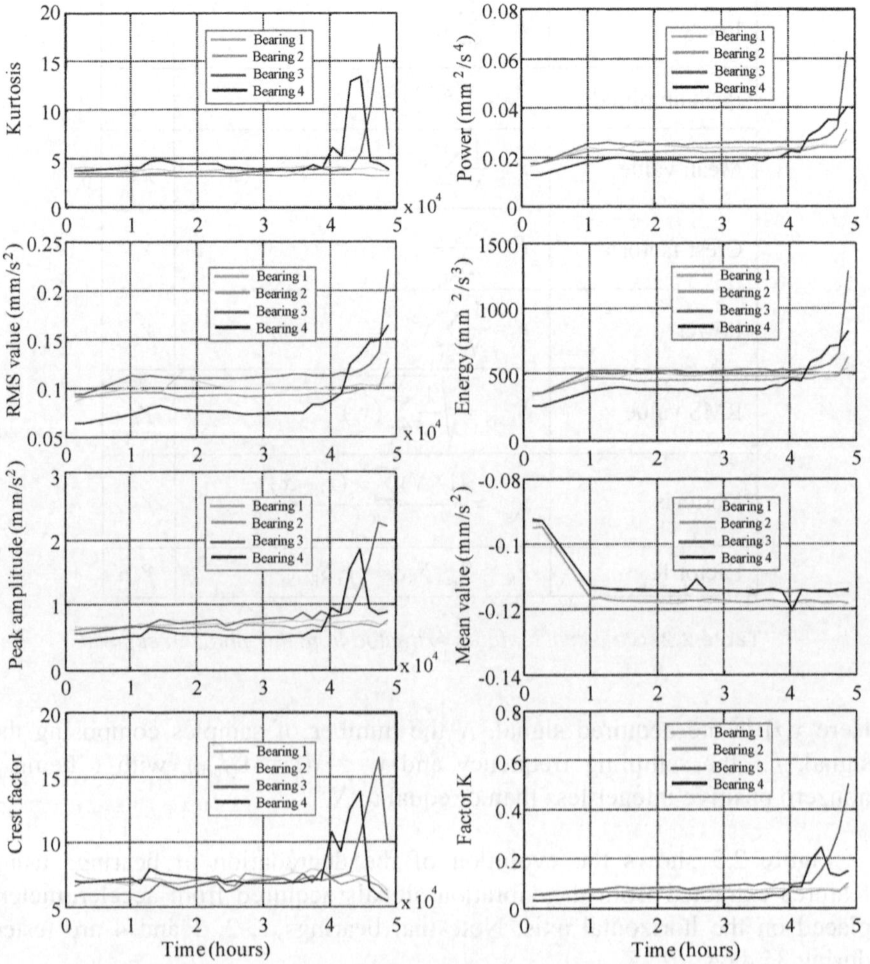

Figure 2.5. *Evolution of the degradation of bearings according to statistical features*

An unsupervised classification algorithm is used to identify the unknown *M*-degradation state from the historical degradation of bearings 1, 2, 3 and 4. For the selected features (γ_3, γ_5, γ_6), several observations disturbed by noise can appear. Thus, a class that groups observations of the same DS can be represented by a geometric zone of observations. If the features are well chosen, each DS can be represented by a compact class grouping similar observations in the representation space as shown in Figure 2.6.

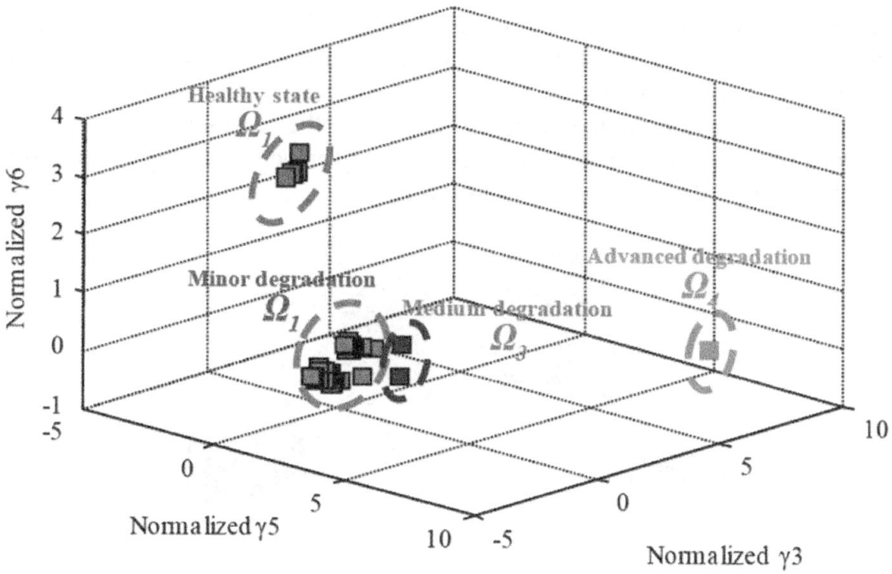

Figure 2.6. *Highlighting the DSs of bearings*

2.1.2. *Degradation of gears*

Figure 2.7 shows the test bench used to observe the degradation of gears and bearings. This latter is comprised of an induction motor, a gear box and an electromagnetic brake. The gear box is comprised of three rotating shafts. The first shaft noted as AE, also called the input shaft, is directly driven by the rotor shaft of the induction motor. The output shaft noted AS is connected to an electromagnetic brake which applies different levels of load. The second shaft noted AI contains the gear and bearing tests (the red zone in Figure 2.7). The induction motor (2.2 kW) is powered by a pulse generator to adjust its speed by varying the rotating frequencies (25, 35 and 45 Hz). The achieved experiments correspond to two load levels (0% and 75%). The acquired signals correspond to the three-phase current signal absorbed by the induction motor. Each current phase is acquired for 10 s with a sample frequency of 25.6 kHz.

In this application, seven experiments were performed on the test bench. These experiments are summarized in Table 2.3.

The proposed features to observe the degradation of gears are expressed by the ratio between the RMS value and standard deviation of the signal. It can be seen from Figure 2.8 that the proposed health indicator allows clearly the separation of seven classes corresponding to the seven experiments in different operating conditions, 45, 35 and 25 Hz with 0% and 75% of load levels. These results highlight the efficiency of the representation space to characterize the different health states of the gear box by using the current signals, even in the cases of low load levels.

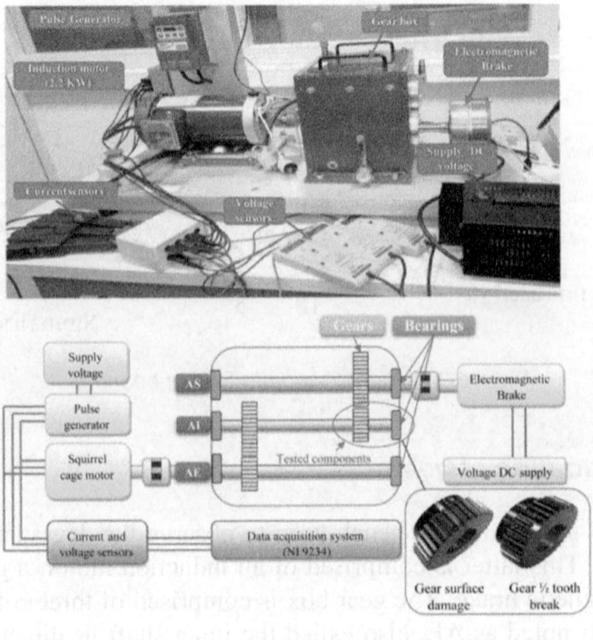

Figure 2.7. *Gear and bearing diagnostic test bench*

	Experiment	Gear	Bearing	Legend
E1	Healthy state	Healthy	Healthy	++
E2	Gear surface damage	Surface damage	Healthy	++
E3	Gear ½ tooth break	½ Tooth break	Healthy	++
E4	Bearing inner race fault	Healthy	Inner race	++
E5	Bearing outer race fault	Healthy	Outer race	
E6	Combined fault (E2 + E4)	Surface damage	Inner race	++
E7	Combined fault (E3 + E5)	½ Tooth break	Outer race	++

Table 2.3. *List of experiments performed on the test bench*

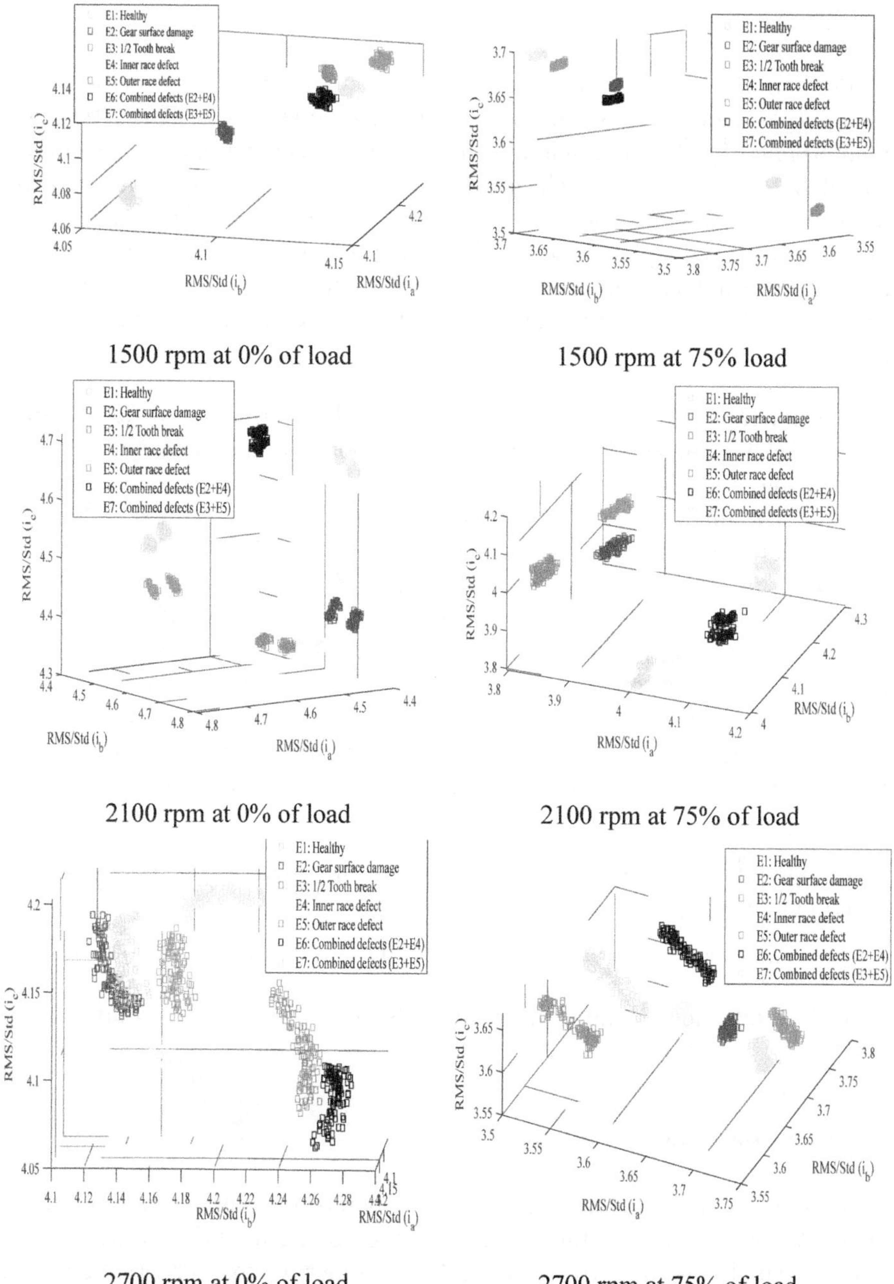

Figure 2.8. *DSs of bearings and gears*

2.2. Prediction

Prediction consists of estimating the evolution of the degradation of a system (gears and bearings) until the appearance of a failure. This estimation is interpreted by the imminence of the failure's detection and its remaining useful life estimation. Baruah and Chinnam [BC 05] showed that HMMs, used in the diagnostics field, can be applied in the field of prognosis. The objective of this chapter is to introduce the prediction methods that are able to detect the defect's imminence of bearings and gears. On this basis, we propose the HMMs as a foundational approach to construct these prognostic methods.

Let Ω_1, Ω_2, ..., Ω_M represent the *M*-classes that define the *M*-degradation states of a system. The first class (Ω_1) corresponds to a system with healthy functioning and the last class (Ω_M) corresponds to a degraded system (faulty). The intermediate classes correspond to a system with several DSs. These *M*-classes, where the number *M* is unknown, will be determined by an unsupervised classification method. Each of these classes occupies a geometric space (representation space) composed by a set of observations projected into a *q*-dimensional space (see Figure 2.2). These observations are collected from the degradation history (from a healthy state until the appearance of a fault) of one or more identical systems. *M*-classes are used to design prognostic methods based on HMMs to detect the imminence of the next stage of degradation and thus the imminence of a system defect. For this, we proposed three methods:

– a method based on the identification of the degradation process of the system under study;

– two methods based on the analysis of the observations sequence (OS).

Before introducing these three methods, let us begin by introducing the HMMs.

2.2.1. *Hidden Markov models*

In a Markov process, each DS corresponds to a physical observable event. More precisely, in the example of Figure 2.2, the DSs are defined by classes which are directly observable. In other words, observations are deterministic functions of the DS. However, in many applications, this approach is too restrictive because DSs are not observable. The Markov process is extended to include the case where the observation is a probabilistic function of the DS. The resulting model is called the HMM or probabilistic function of the Markov chains [RAB 89].

HMM is a double stochastic process, with an underlying stochastic process (DS sequence) that is not directly observable but can only be observed through another stochastic process that results in an OS.

2.2.1.1. *Parameters of an HMM*

An HMM is defined by a set of finite DSs connected by transitions (see Figure 2.9). Each state is characterized by two sets of probabilities: (i) a first-order Markov chain transition probability and (ii) a discrete output probability that defines the probability of observing an observation, given the system DS. The elements of a HMM can be defined as follows:

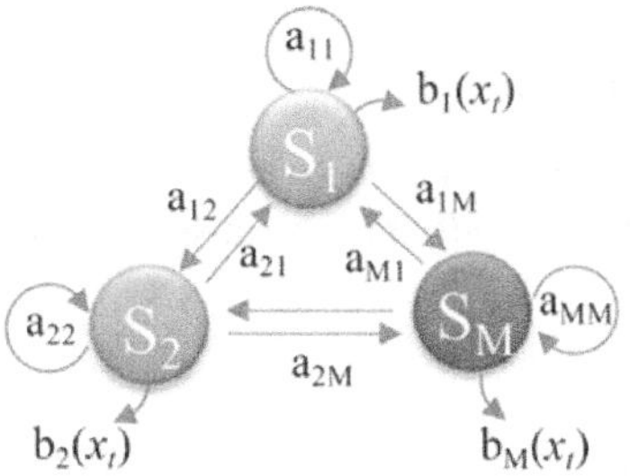

Figure 2.9. *Schematic representation of an HMM*

Number of DSs

The number of DSs in an HMM is denoted M. A set of hidden M-degradation states is defined by $S = \{S_1, S_2, ..., S_M\}$. The DS at time t is given by $q_t \in S$, $0 \leq t \leq T$, where T is the length of the observation sequence and q_t is the current DS.

Transition probability

The transition probability matrix is given by $A = \{a_{ij}\}$ with $1 \leq i, j \leq M$. a_{ij} represents the transition probability from a DS (S_i) observed at time $t - 1$ to another DS (S_j) observed at the time instant t. The transition probability is defined as

$$a_{ij} = P(q_t = S_j \mid q_{t-1} = S_i), \quad 1 \leq i, j \leq M. \qquad [2.1]$$

The transition probability must satisfy the following constraints:

$$a_{ij} \geq 0, \quad 1 \leq i, j \leq M,$$
$$\sum_{j=1}^{M} a_{ij} = 1.$$

Probability of observation

An observation probability matrix denoted B (B=$b_j(x_t)$) indicates the probability of observing an observation x_t associated with a given DS (S_j) at time t. This probability is defined as

$$b_j(x_t) = P(x_t \mid q_t = S_j), \quad 1 \leq j \leq M, \quad 0 \leq t \leq T, \qquad [2.2]$$

where $0 \leq b_j(x_t) \leq 1$, x_t designates an observation at time t.

Vector of initial probabilities

The initial probability vector $\pi = \{\pi_i\}_{i\in(1,\dots,M)}$, indicates the probability of starting in a DS (S_i) at time $t = 0$. This vector is expressed as

$$\pi_i = P(q_0 = S_i), \quad 1 \leq i \leq M, \qquad [2.3]$$

where $0 \leq \pi_i \leq 1$ and $\sum_{i=1}^{M} \pi_i = 1$.

The following notation is often used in the literature to represent an HMM:

$$\lambda = (A, B, \pi).$$

2.2.2. *Fault prediction by identifying the degradation process of the system*

The first proposed "predictor" is based on modeling the degradation process of N-identical systems. The aim is to predict the imminence of the next DS of a new, similar system by using the HMMs modeled from the degradation processes of N-systems. In this case, the degradation process of each system k is modeled by an HMM λ_k, k∈(1,2, ..., N). The parameters of a model (λ_k) are determined by learning an OS noted $X^k = \{x_0^k, x_1^k, x_2^k, \dots, x_T^k\}$ representing the degradation process of a system k up to the instant T.

The learning of an HMM $\lambda = (A, B, \pi)$ from a sequence of observations X^k is obtained by the Baum–Welch re-estimation formulas. The specification of the three sets of probability A, B and π is then necessary in order to have a complete description of each degradation process of the system. In the following, the notation used to describe an HMM to represent a degradation process k of the system is given by

$$\underset{1\le k\le N}{\lambda_k} = (A^{(k)}, B^{(k)}, \pi^{(k)}) \quad [2.4]$$

with

$$A^{(k)} = \{a_{ij}^k\}, \quad 1 \le i, j \le M, \quad 1 \le k \le N, \quad [2.5]$$

$$\pi^{(k)} = \{\pi_i\}, \quad 1 \le i \le M, \quad [2.6]$$

$$B^{(k)} = \underset{\substack{1\le j\le M \\ 1\le k\le N}}{b_j(X^k)} = 1 - \frac{(X^k - g_j)\mathrm{Cov}^{-1}(E_j)(X^k - g_j)'}{\sum_{\ell=1}^{M}(X^k - g_\ell)\mathrm{Cov}^{-1}(E_\ell)(X^k - g_\ell)'}, \quad [2.7]$$

where k corresponds to the kth system, E_j groups the learning observations of the class Ω_j, g_j is the center of gravity of the class Ω_j and $(\cdot)'$ denotes a transposed matrix. The covariance matrix Cov is estimated from each class.

2.2.2.1. *Learning the degradation process of a system by the HMM*

The only available information about the system is its different DSs (its classes). Thus, the proposed approach consists of modeling the degradation process of the system using an HMM from an OS X^k of length T, $\underset{1\le k\le N}{X^k} = \{x_0^k, x_1^k, x_2^k, \ldots, x_T^k\}$ (see Figure 2.10). This sequence corresponds to a successive draw of observations up to the instant T of the system k. This OS is used to adjust the parameters of each HMM to maximize the probability $P(X^k|\lambda_k)$corresponding to the probability of generating the OS X^k given the degradation process model λ_k.

As we can see in Figure 2.10, the observations composing the sequence X^k evolve over time since $x_0^k \neq x_1^k \neq x_2^k \neq \cdots \neq x_T^k$. This implies a high probability of transition from a DS to another. If q_{t-1} designates the DS (S_i) where the observation is placed at time $t-1$, then transition to the next DS ($q_t = S_j$) can be expressed as:

$$a_{ij} = P(q_t = S_j \mid q_{t-1} = S_i, \lambda_k), \quad 1 \le k \le N, \quad 1 \le i, j \le M. \qquad [2.8]$$

This transition probability is calculated by the Baum–Welch re-estimation formula.

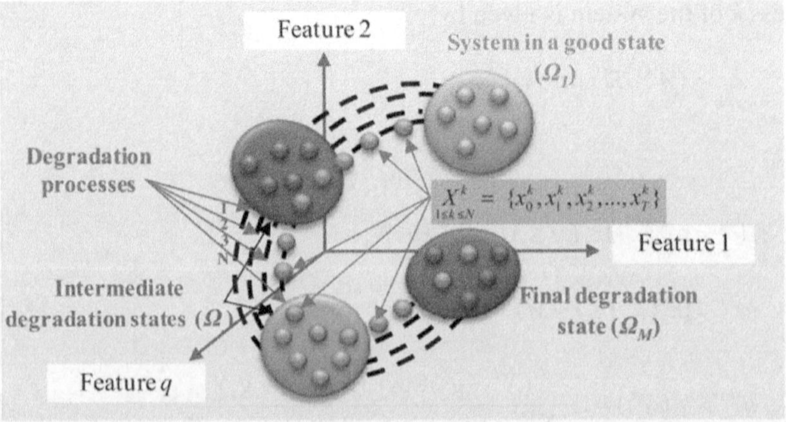

Figure 2.10. *Extraction of an OS of a degradation process*

The Baum–Welsh re-estimation algorithm starts from an initial model (λ_k) and defines an iterative algorithm to maximize the probability $P(X^k|\lambda_k)$ in order to estimate the most representative model of the OS X^k. Implementation of the Baum–Welch algorithm involves calculation of two different probabilities. First, the variable "forward" which is defined by the probability of a partial sequence of observations $\{x_0^k, x_1^k, x_2^k, ..., x_t^k\}$ extracted from the degradation process k ending in the DS (S_i) at time t.

2.2.2.1.1. The forward variable

The forward variable denoted $\alpha_t^{(k)}(i)$ is defined by

$$\alpha_t^{(k)}(i) = P(x_0^k, x_1^k, x_2^k, ..., x_t^k, q_t = S_i \mid \lambda_k), \quad 1 \le i \le M, \quad 1 \le k \le N, \quad 0 \le t \le T. \qquad [2.9]$$

The forward algorithm is determined as follows:

1. Initialization:

$$\alpha_0^{(k)}(i) = \pi_i^{(k)} b_i(x_0^k). \qquad [2.10]$$

2. Recurrence:

$$\alpha_t^{(k)}(j) = b_j(x_t^k)\sum_{i=1}^{M}\alpha_{t-1}^{(k)}(i)a_{ij}^{(k)}, \quad 1 \le j \le M, \quad 1 \le k \le N, \quad 1 \le t \le T \qquad [2.11]$$

with $b_j(x_t^k) = P(x_t^k \mid q_t = S_j)$ being the probability of observing an observation x_t extracted from the system k associated with the DS (S_j) at time t and $a_{ij}^{(k)}$ transition probability from the DS S_i at time t–1 to another DS S_j at time t given the model λ_k of the degradation process.

2.2.2.1.2. The backward variable

The backward variable denoted $\beta_t^{(k)}(i)$ is defined by

$$\beta_t^{(k)}(i) = P\left(x_{t+1}^k, \ldots, x_T^k \mid q_t = S_i, \lambda_k\right), \quad 1 \le i \le M, \quad 1 \le k \le N, \quad 0 \le t \le T. \qquad [2.12]$$

The backward algorithm is determined as follows:

1. Initialization:

$$\beta_T^{(k)}(i) = 1. \qquad [2.13]$$

2. Recurrence:

$$\beta_t^{(k)}(i) = \sum_{j=1}^{M}\beta_{t+1}^{(k)}(j)b_j(x_{t+1}^k)a_{ij}^{(k)}, \quad 1 \le i \le M, \quad 1 \le k \le N, \quad 0 \le t \le T-1. \qquad [2.14]$$

The goal is to maximize the probability $P(X^k|\lambda_k)$. This requires using the forward and backward variables, as well as the probability of transition from the DS S_i at time t to the DS S_j at time t+1 given the OS X^k and the initial model λ_k to estimate the expectation of transition from S_i to S_j. This expectation is denoted $\xi_t^{(k)}(i,j) = P(q_{t+1} = S_j \mid q_t = S_i, X^k, \lambda_k)$ and is expressed as

$$\xi_t^{(k)}(i,j) = \frac{P(q_t = S_i, q_{t+1} = S_j, X^k \mid \lambda_k 0}{P(X^k \mid \lambda_k)}. \qquad [2.15]$$

By using the forward and backward variables, we obtain

$$\xi_t^{(k)}(i,j) = \frac{\alpha_t^{(k)}(i) a_{ij}^{(k)} b_j(x_{t+1}^k) \beta_{t+1}^{(k)}(j)}{\sum_{i=1}^{M} \alpha_t^{(k)}(i) \beta_t^{(k)}(i)}. \quad [2.16]$$

Consider also the variable $\gamma_t^{(k)}(i)$:

$$\gamma_t^{(k)}(i) = P(q_t = S_i \mid X^k, \lambda_k). \quad [2.17]$$

The variable $\gamma_t^{(k)}(i)$ expresses the probability of being in a DS S_i at time t given the OS X^k and the model λ_k.

Expressed according to $\xi_t^{(k)}(i,j)$, $\gamma_t^{(k)}(i)$ gives

$$\gamma_t^{(k)}(i) = \sum_{j=1}^{M} \xi_t^{(k)}(i,j). \quad [2.18]$$

If we sum $\gamma_t^{(k)}(i)$ over all the instants t, we obtain a quantity interpreted as the expectation of the number of times that the DS S_i is used to generate the sequence X^k. Similarly, if we sum $\xi_t^{(k)}(i,j)$ on all the instants t, we obtain a quantity that can be interpreted as the expectation of the number of times that the transition $S_i \rightarrow S_j$ is used to generate the sequence X^k. So, we have

$$\sum_{t=0}^{T-1} \gamma_t^{(k)}(i) = \text{expectation of the number of transitions to } S_i, \quad [2.19]$$

$$\sum_{t=0}^{T-1} \xi_t^{(k)}(i,j) = \text{expectation of the number of transitions from } S_i \text{ to } S_j. \quad [2.20]$$

By combining the two previous formulas, we can then propose a set of parameter estimators:

$$\pi_i^{(k)} = \gamma_0^{(k)}(i), \quad [2.21]$$

$$a_{ij}^{(k)} = \frac{\sum_{t=0}^{T-1} \xi_t^{(k)}(i,j)}{\sum_{t=0}^{T-1} \gamma_t^{(k)}(i)}, \quad 1 \le i,j \le M, \quad 1 \le k \le N. \quad [2.22]$$

Once the update is complete, we obtain a set of new parameters for each model $\underset{1\le k\le N}{\lambda_k} \rightarrow (A^{(k)},\pi^{(k)})$. The Baum–Welch algorithm can be summarized as follows:

(1) Input data $X^k = \{x_0^k, x_1^k, x_2^k, \ldots, x_T^k\}$
(2) Choose an initial setting of $(A^{(k)},\pi^{(k)})$
(3) **Repeat**
(4) **For** each $\underset{0\le t\le T}{x_t^k} \in X^k$
(5) Calculate the values of the variables $\alpha_t^{(k)}(i)$ using the forward algorithm [2.9]
(6) Calculate the values of the variables $\beta_t^{(k)}(i)$ using the backward algorithm [2.12]
(7) Calculate the values of the variables $\xi_t^{(k)}(i,j)$ [2.16]
(8) Calculate the values of the variables $\gamma_t^{(k)}(i)$ [2.18]
(9) Calculate and store expression values of [2.19] and [2.20]
(10) **End For**
(11) Re-estimate the parameters $(A^{(k)},\pi^{(k)})$ using the formulas [2.21] and [2.22]
(12) **Until** a stop criterion.

Table 2.4. *Baum–Welch learning algorithm*

For a new OS $X = \{x_0, x_1, x_2, \ldots, x_t\}$ measured at regular time intervals from a new system, if $q_{t+1} = S_{j\ j\in(1,\ldots,\ M)}$, denotes the DS of a system at time t+1, and if the DS of the system $q_{t+1} = S_j$ is different than $q_t = S_i$, then we can predict the imminence of the next DS by using formula [2.24].

The first step consists of identifying the HMM λ_k, $k\in(1,2,\ldots,N)$, representing the best new OS X. This is assimilated as a similarity measure given by the probability of generating the sequence X given the model λ_k. The model for which this probability is maximum will define the most similar degradation process. This selection given by

$$\lambda_k = \underset{1\le k\le N}{\arg\max}[P(x_0, x_1, \ldots, x_t \mid \lambda_k)] = \underset{1\le k\le N}{\arg\max}\left[\sum_{i=1}^{M} \alpha_t^{(k)}(i)\beta_t^{(k)}(i)\right]. \qquad [2.23]$$

The model identification calculates the probability that the OS $\{x_0, x_1, x_2, \ldots, x_t, x_{t+1}\}$ arrives at the state S_j at time t+1 possible, given the model λ_k. This is interpreted by the following expression: $P(x_0, x_1, x_2, \ldots, x_t, x_{t+1}, q_{t+1} = S_j \mid \lambda_k)$, with x_{t+1} unknown. This expression allows determining among the M-degradation states S_j, $j\in(1,...,M)$, the one that maximizes this probability. If the future DS $q_{t+1}=S_j$ is different from the actual DS at time t, that is, $q_t = S_i$, then we can predict the imminence of the next stage of degradation using the following formula:

$$\text{Degradation state } S_j \text{ at } (t+1) = \underset{\substack{1\le j\le M\\ 1\le k\le N}}{\arg\max}(P(x_0, x_1, x_2, \ldots, x_t, x_{t+1}, q_{t+1} = S_j \mid \lambda_k)). \quad [2.24]$$

An algorithm is proposed to predict the DS of the system at t+1. For this, we define the variable $\delta_t(i)$ as the normalized forward probability of the DS S_i at time t.

The DS S_j at t+1 is given by

$$\text{Degradation state } S_j \text{ of the system at } (t+1) = \underset{\substack{1\le j\le M\\ 1\le k\le N}}{\arg\max}\left(\sum_{i=1}^{M}\delta_t^{(k)}(i)a_{ij}^{(k)}\right) \quad [2.25]$$

with $\delta_{t+1}^{(k)}(i) = b_j(x_{t+1})\alpha_t^{(k)}(i)$.

2.2.2.1.3. Proof

According to the forward and backward formulations, the probability that an OS $X = \{x_0,..., x_T\}$ evolves until the instant T given the model λ_k is given by

$$\begin{aligned} P(x_0, x_1, x_2, \ldots, x_T \mid \lambda_k) &= \sum_{j=1}^{M} P(x_0, x_1, x_2, \ldots, x_t, q_t = s_j \mid \lambda_k) P(x_{t+1}, \ldots, x_T \mid q_t = s_j, \lambda_k) \\ &= \sum_{j=1}^{M} \alpha_t^{(k)}(j)\beta_t^{(k)}(j), \quad 1\le j\le M, \quad 1\le k\le N. \end{aligned} \quad [2.26]$$

Formula [2.26] allows deducing that for an OS of length $T = t+1$, the backward variable $\beta_{t+1}^{(k)}(j) = 1$. This conclusion allows us to rewrite [2.26] in the following form:

$$P(x_0, x_1, x_2, \ldots, x_t, x_{t+1}, q_{t+1} = S_j \mid \lambda_k) = \alpha_{t+1}^{(k)}(j) = b_j(x_{t+1})\sum_{i=1}^{M}\alpha_t^{(k)}(i)a_{ij}^{(k)}. \quad [2.27]$$

From formula [2.27], the probability that the OS $\{x_0, x_1, x_2, \ldots, x_t, x_{t+1}\}$ arrives to the DS S_j at t+1 depends on the transition probability $a_{ij}^{(k)}$, the forward variable $\alpha_t^{(k)}(i)$ at the time instant t and the observation's emission probability at t+1 given the state S_j, $b_j(x_{t+1})$.

On the basis of formula [2.27], the next DS can be deduced by finding the state S_j that maximizes the previous formula. Therefore, the next DS is expressed as

$$\text{Degradation state } S_j \text{ of the system at } (t+1) = \underset{\substack{1\leq j\leq M \\ 1\leq k\leq N}}{\arg\max}\left(b_j(x_{t+1})\sum_{i=1}^{M}\alpha_t^{(k)}(i)a_{ij}^{(k)} \right). \quad [2.28]$$

As we can see from formula [2.28], the observation's emission probability $b_j(x_{t+1})$ is unknown. This probability is between "0" and "1." A probability of 0 means that there is a low probability of observing x_{t+1} associated with the DS S_j and the opposite is true for a probability of 0. The objective is to detect the imminence of the next DS. This estimation is obtained by supposing a high observation emission probability which results in the normalization of $b_j(x_{t+1})$. This normalization is interpreted by the fact that the probability of observing an observation at t+1 is maximum for all the possible states of degradation S_j of the system. Thus, we obtain

$$P(x_0, x_1, x_2, \ldots, x_t, x_{t+1}, q_{t+1} = S_j \mid \lambda_k) = \sum_{i=1}^{M}\delta_t^{(k)}(i)a_{ij}^{(k)} \quad [2.29]$$

with $\delta_0^{(k)}(i) = \pi_i^{(k)}b_i(x_0)$ and $\delta_{t+1}^{(k)}(j) = b_j(x_{t+1})\sum_{i=1}^{M}\delta_t^{(k)}(i)a_{ij}^{(k)}$, hence the formula [2.25].

The following shows the steps to follow to predicting the imminence of the next DS:

1. Initialization:

$$\delta_0^{(k)}(i) = \pi_i b_i(x_0).$$

2. Recurrence:

$$\text{Degradation state } S_j \text{ of the system at } (t+1) = \underset{1\leq j\leq M}{\arg\max}\left(\sum_{i=1}^{M}\delta_t^{(k)}(i)a_{ij}^{(k)} \right),$$

$$\delta_{t+1}^{(k)}(j) = b_j(x_{t+1})\sum_{i=1}^{M}\delta_t^{(k)}(i)a_{ij}^{(k)}, \quad 1\leq i, j\leq M, \quad 1\leq k\leq N, \quad t\geq 0.$$

Figure 2.11 shows the flowchart of the first method based on the identification of the degradation process.

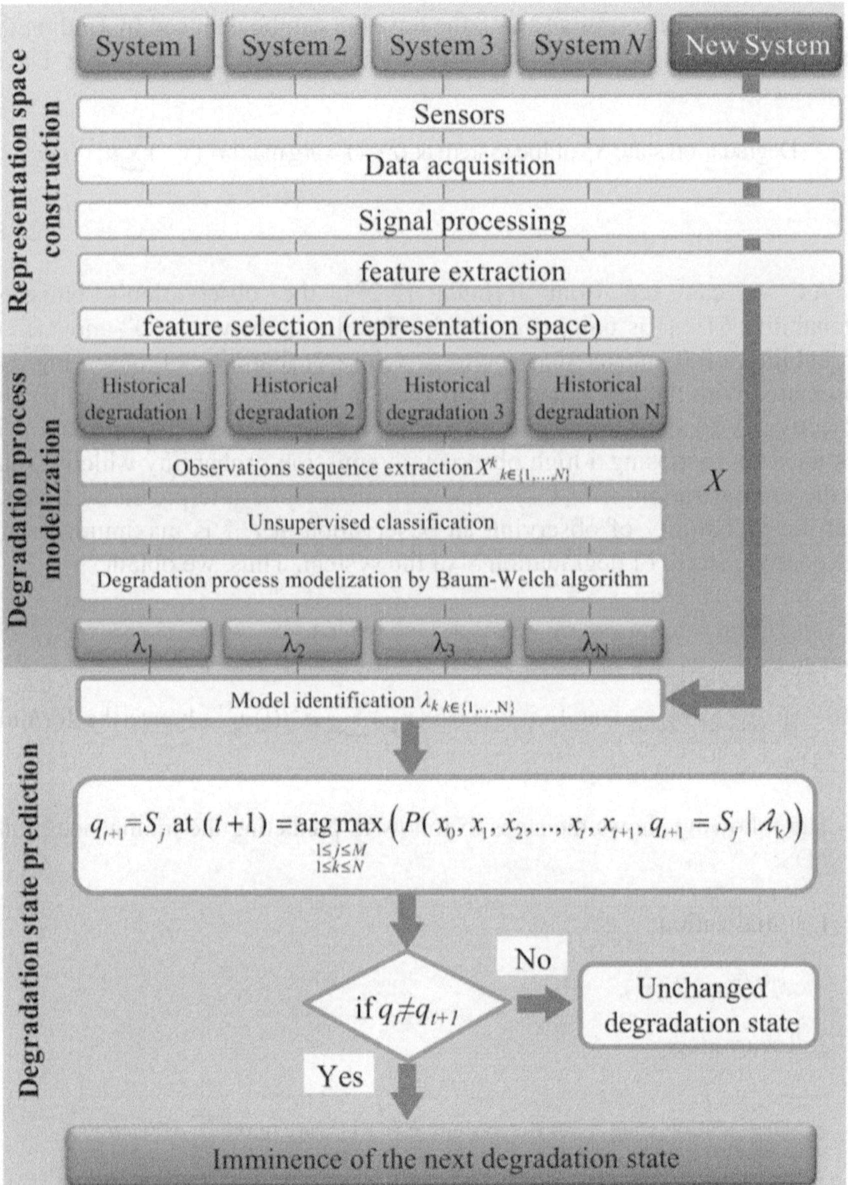

Figure 2.11. *Flowchart of the first prognosis method based on the identification of the degradation process*

2.2.2.1.4. Illustration

Here, we illustrate the applicability of the first method on the historic degradation of 10 identical bearings placed on a production line for a duration t estimated in months. The representation space is composed by a set of q-features indicating the degradation of bearings. Measurements made on bearings can distinguish three DSs: minor DS, medium DS and an advanced state of degradation (see Figure 2.12). We deduce that the degradation of each motor passes systematically through these three DS as shown in Figure 2.12.

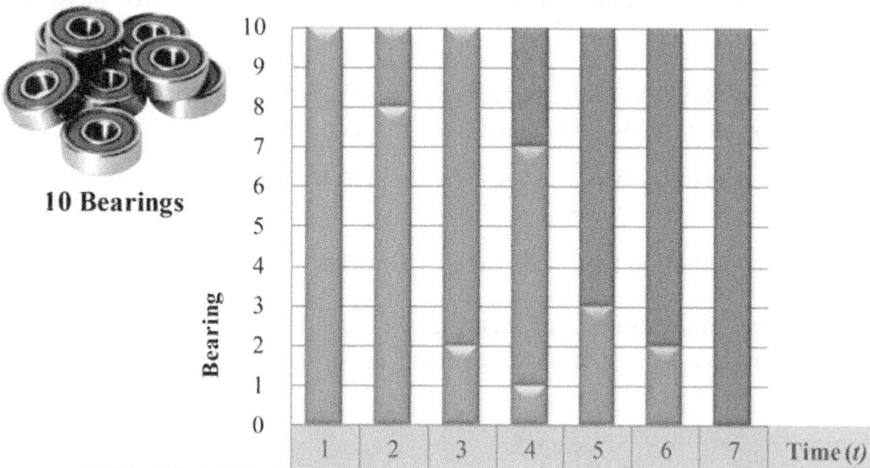

	1	2	3	4	5	6	7
Advanced DS (3)	0	0	0	3	7	8	10
Medium DS (2)	0	2	8	6	3	2	0
Minor DS (1)	10	8	2	1	0	0	0

DS : degradation state

Figure 2.12. *Degradation evolution of 10 bearings*

It can be seen from Figure 2.12 that each DS corresponds to a set of observations projected in a q-dimensional space. The degradation process starts by the minor degradation and then the medium degradation arrives in the advanced DS. The OS extracted from each motor is formulated as follows:

$$X^k = \{x_0^k, x_1^k, x_2^k, \ldots, x_t^k\}, \quad 0 \leq t \leq 7, \quad 1 \leq k \leq 10$$

with $x_t^k = [x_{t,1}^k, x_{t,2}^k, x_{t,3}^k, \ldots, x_{t,q}^k]$ which designates an observation measured on the bearing k at the time instant t described by the feature vector.

Consider a new OS extracted from measurements made on a new bearing until time t=6 months, numbered 11, identical to the previous ones $X^{11} = \{x_0^{11}, x_1^{11}, x_2^{11}, x_3^{11}, x_4^{11}, x_5^{11}, x_6^{11}\}$. The goal of the predictor is to detect the imminence of the next DS at time t+1 = 7.

To predict the imminence of the next DS on a new bearing, we first need to build an HMM from the degradation process for each bearing by using their degradation history. For this, we need to have the following:

(1) Define the DSs. Note that $S = \{S_1, S_2, S_3\}$ corresponds to a set of three DSs: minor, medium and advanced DSs, respectively.

(2) The transition between the DSs defined by the transition matrix A of dimension (3×3).

(3) An observation matrix B given by formula [2.7].

Consider the example of bearing 1 indexed k=1 and in which its degradation process is similar to bearing 1 as shown in Figure 2.13.

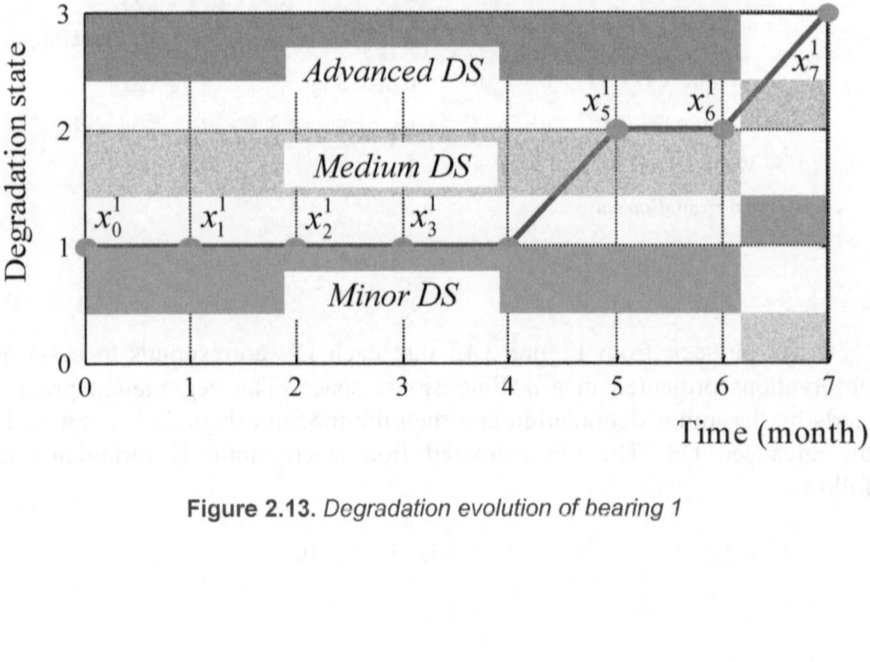

Figure 2.13. *Degradation evolution of bearing 1*

The OS extracted from bearing 1 until the moment t=6 is formulated as follows: $X^1 = \{x_0^1, x_1^1, x_2^1, x_3^1, x_4^1, x_5^1, x_6^1\}$.

The observation matrix B given by (3.10) is summarized in Table 2.5:

$$\underset{\substack{1\leq j\leq 3\\ k=1\\ 0\leq t\leq 6}}{b_j(x_t^k) = P(x_t^k \mid q_t = S_j)} = 1 - \frac{(x_t^k - g_j)\mathrm{Cov}^{-1}(E_j)(x_t^k - g_j)'}{\sum_{\ell=1}^{M}(x_t^k - g_\ell)\mathrm{Cov}^{-1}(E_\ell)(x_t^k - g_\ell)'}. \qquad [2.30]$$

where E_j represents the group of observations contained in the jth Ω_j, g_j is the gravity center the class Ω_j and (.)′ denotes a transposed matrix. The covariance matrix Cov is estimated from each class.

Motor 1 (k = 11)	x_0^1	x_1^1	x_2^1	x_3^1	x_4^1	x_5^1	x_6^1
Minor degradation $b_1(x_t^1) = P(x_t^1 \mid q_t = S_1)$	0.8	0.7	0.6	0.5	0.5	0.3	0.1
Medium degradation $b_1(x_t^1) = P(x_t^1 \mid q_t = S_2)$	0.2	0.2	0.3	0.4	0.4	0.5	0.5
Major degradation $b_1(x_t^1) = P(x_t^1 \mid q_t = S_3)$	0	0.1	0.1	0.1	0.1	0.2	0.4

Table 2.5. *Observation matrix B corresponding to the OS X^1*

The initial parameters ($A^{(1)}$, $\pi^{(1)}$) of the degradation process of motor 1 (λ_1) are given as

$$A^{(1)} = \begin{matrix} & S_1 & S_2 & S_3 \\ S_1 & 0.5 & 0.25 & 0.25 \\ S_2 & 0 & 0.5 & 0.5 \\ S_3 & 0 & 0 & 1 \end{matrix}, \quad \pi^{(1)} = \begin{matrix} S_1 & S_2 & S_3 \\ \frac{1}{3} & \frac{1}{3} & \frac{1}{3} \end{matrix}.$$

Note that the transition probabilities a_{32}, a_{31}, a_{21} are set to zero. In our case, class Ω_1 corresponds to a healthy system, and Ω_2 and Ω_3 correspond to progressive degradation of the system. In this case, it is normal to set a_{32}, a_{31} and a_{21} to zero because we consider that a degraded system cannot return to normal operation.

From the forward and backward formulations [2.21] and [2.22], we obtain a re-estimation of the transition matrix $A^{(1)}$ and the initial state vector $\pi^{(1)}$:

$$A^{(1)} = \begin{matrix} & S_1 & S_2 & S_3 \\ S_1 & 0.7 & 0.2 & 0.1 \\ S_2 & 0 & 0.65 & 0.35 \\ S_3 & 0 & 0 & 1 \end{matrix}, \quad \pi^{(1)} = \begin{matrix} S_1 & S_2 & S_3 \\ (0,7 & 0,2 & 0,1) \end{matrix}.$$

To apply the prognostic method, we first need to identify the HMM λ_k, $1 \leq k \leq N$, for which the observation probability of the sequence X^{11} is maximal. The identification of the HMM is obtained by

$$\lambda_k = \underset{1\leq k\leq 10}{\arg\max}[P(x_0^{11}, x_1^{11}, x_2^{11}, x_3^{11}, x_4^{11}, x_5^{11}, x_6^{11} \mid \lambda_k)] = \underset{1\leq k\leq 10}{\arg\max}\left[\sum_{i=1}^{M} \alpha_t^{(k)}(i)\beta_t^{(k)}(i)\right]. \quad [2.31]$$

The observation matrix B corresponding to this new OS X^{11} is given in Table 2.6.

Motor 1 ($k=11$)	x_0^{11}	x_1^{11}	x_2^{11}	x_3^{11}	x_4^{11}	x_5^{11}	x_6^{11}
Minor degradation $b_1(x_t^{11}) = P(x_t^{11}\ q_t = S_1)$	0.75	0.6	0.5	0.45	0.35	0.3	0.1
Medium degradation $b_1(x_t^{11}) = P(x_t^{11} \mid q_t = S_2)$	0.2	0.3	0.35	0.4	0.45	0.5	0.47
Major degradation $b_1(x_t^{11}) = P(x_t^{11} \mid q_t = S_2)$	0.05	0.1	0.15	0.15	0.2	0.2	0.43

Table 2.6. *Observation matrix B corresponding to the OS X^{11}*

The degradation process λ_1 of bearing 1 is the most representative of the OS X^{11} with a probability $P(X^{11}|\lambda_1)$ equals to 1.63×10^{-3}. This similarity allows to predict the imminence of the next DS at $t + 1=7$ by calculating the probability that the OS of X^{11} arrives to the DS S_j at $t + 1$, given λ_1. The degradation process model λ_1 of bearing 1 used to find among the M-degradation states $S_j, j \in (1,\ldots,M)$, the one which maximizes the formula

$$P(x_0^{11}, x_1^{11}, x_2^{11}, x_3^{11}, x_4^{11}, x_5^{11}, x_6^{11}, x_{t+1}^{11}, q_{t+1} = S_j \mid \lambda_1) = \sum_{i=1}^{M} \delta_t^{(1)}(i) a_{ij}^{(1)}. \quad [2.32]$$

Table 2.7 summarizes the normalized forward variables obtained for each DS up to time $t = 6$.

	Minor degradation	Medium degradation	Advanced degradation
$t=0$	$\delta_0^1(1)=0.525$	$\delta_0^1(2)=0.04$	$\delta_0^1(3)=5\times10^{-3}$
$t=1$	$\delta_1^1(1)=0.2205$	$\delta_1^1(2)=0.0393$	$\delta_1^1(3)=7.15\times10^{-3}$
$t=2$	$\delta_2^1(1)=0.077$	$\delta_2^1(2)=0.0243$	$\delta_2^1(3)=6.44\times10^{-3}$
$t=3$	$\delta_3^1(1)=0.0242$	$\delta_3^1(2)=0.0125$	$\delta_3^1(3)=2.26\times10^{-3}$
$t=4$	$\delta_4^1(1)=5.9\times10^{-3}$	$\delta_4^1(2)=5.83\times10^{-3}$	$\delta_4^1(3)=1.18\times10^{-3}$
$t=5$	$\delta_5^1(1)=1.23\times10^{-3}$	$\delta_5^1(2)=2.48\times10^{-3}$	$\delta_5^1(3)=7.012\times10^{-4}$
$t=6$	$\delta_6^1(1)=8.61\times10^{-5}$	$\delta_6^1(2)=8.73\times10^{-4}$	$\delta_6^1(3)=8.23\times10^{-4}$

Table 2.7. *Normalized forward variables obtained from the OS X^{11} up to time t = 6*

For $t+1=7$, we have the following:

$$S_j(t+1)=\arg\max\left[\sum_{i=1}^{M}\delta_t^{(1)}(i)a_{ij}^{(1)}\right]\Leftrightarrow S_j(t+1=7)=\arg\max_{1\le j\le M}\left[\sum_{i=1}^{M}\delta_6^{(1)}(i)a_{ij}^{(1)}\right]$$

$$\begin{aligned}\text{for}\quad q_{t+1}=S_1 &\Rightarrow \sum_{i=1}^{M}\delta_6^{(1)}(i)a_{i1}^{(1)}=\delta_6^{(1)}(1)a_{11}^{(1)}+\delta_6^{(1)}(2)a_{21}^{(1)}+\delta_6^{(1)}(3)a_{31}^{(1)}\\ &=(8.61\times10^{-5})\times0.7+(8.73\times10^{-4})\times0+(8.32\times10^{-4})\times0\\ &=6.02\times10^{-5}\end{aligned}$$

$$\begin{aligned}\text{for}\quad q_{t+1}=S_2 &\Rightarrow \sum_{i=1}^{M}\delta_6^{(1)}(i)a_{i2}^{(1)}=\delta_6^{(1)}(1)a_{12}^{(1)}+\delta_6^{(1)}(2)a_{22}^{(1)}+\delta_6^{(1)}(3)a_{32}^{(1)}\\ &=(8.61\times10^{-5})\times0.2+(8.73\times10^{-4})\times0.65+(8.32\times10^{-4})\times0\\ &=5.84\times10^{-4}\end{aligned}$$

$$\begin{aligned}\text{for}\quad q_{t+1}=S_3 &\Rightarrow \sum_{i=1}^{M}\delta_6^{(1)}(i)a_{i3}^{(1)}=\delta_6^{(1)}(1)a_{13}^{(1)}+\delta_6^{(1)}(2)a_{23}^{(1)}+\delta_6^{(1)}(3)a_{33}^{(1)}\\ &=(8.61\times10^{-5})\times0.1+(8.73\times10^{-4})\times0.35+(8.32\times10^{-4})\times1\\ &=1.14\times10^{-3}\end{aligned}$$

$$S_j(t+1=7)=\arg\max_{1\le j\le M}\left[\sum_{i=1}^{M}\delta_6^{(1)}(i)a_{ij}^{(1)}\right]=\arg\max\left\{6.02\times10^{-5}\quad 5.84\times10^{-4}\quad 1.14\times10^{-3}\right\}$$

$$\rightarrow q_{t+1}=S_3.$$

The probability that the OS X^{11} arrives at the state S_3 at the time instant $t+1=7$, given the degradation process of motor 1, is greater than that of the stages S_1 and S_2. This allows us to predict the imminence of the third stage of degradation before it occurs (see Figure 2.14).

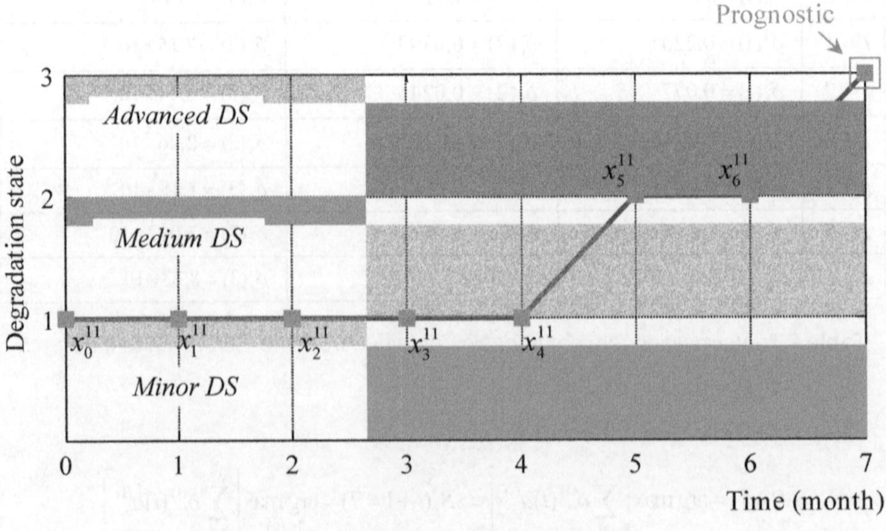

Figure 2.14. *Degradation evolution of motor 11*

2.2.2.2. *Methods based on analysis of the OS*

The purpose of the first method is to detect the imminence of the system's next DS based on the historic degradation process of other systems. The applicability of this method is valid only in the case where the degradation process of the tested system is similar to those studied. However, the availability of such a database is, in most applications, difficult to obtain or impossible. To overcome this problem, we propose two other methods based on the analysis of OS to detect the imminence of the future DS of the system. These two methods are based on the modelization of the DSs and not the degradation process of the system. To find the DSs, it sufficient to have the historic degradation of a single system.

2.2.2.3. *Evolutionary Markov model*

The second method calculates the probability that the OS $X=\{x_0,x_1,x_2,\ldots,x_t\}$ arrives at the DS S_j at time t+1, given an HMM noted

λ_t. This probability is expressed as $P(x_0, x_1, x_2, \ldots, x_t, x_{t+1}, q_{t+1} = S_j \mid \lambda_t)$, with x_{t+1} unknown. The difference compared to the first method is in the construction of the HMM λ_t at time t. Here λ_t is obtained from the OS X^t itself. This OS is used to re-estimate at each time step the HMM corresponding to each new OS. This approach is illustrated in Figure 2.15.

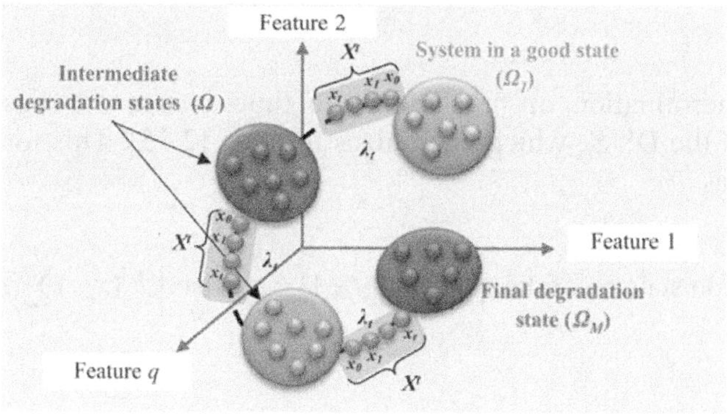

Figure 2.15. *Construction of an evolutionary Markov model*

The imminence of the next DS $q_{t+1} = S_j, j \in (1, \ldots, M)$, is obtained by calculating the probability $P(x_0, x_1, x_2, \ldots, x_t, x_{t+1}, q_{t+1} = S_j \mid \lambda_k)$ and the determination of the stage S_j which maximizes this probability [3.13].

$$
\begin{aligned}
\text{Degradation state } (S_j) \text{ at } (t+1) &= \underset{1 \le j \le M}{\arg\max}(P(x_0, x_1, x_2, \ldots, x_t, x_{t+1}, q_{t+1} = S_j \mid \lambda_t)) \\
&= \underset{1 \le j \le M}{\arg\max}\left(\sum_{i=1}^{M} \delta_t(i) a_{ij}^{(t)} \right).
\end{aligned}
\quad [2.33]
$$

2.2.2.3.1. Proof

According to the forward and backward formulations, the probability that an OS $X^t = \{x_0, \ldots, x_t\}$ evolves to the DS S_j at the time instant t is given as

$$
\begin{aligned}
P(x_0, x_1, x_2, \ldots, x_t \mid \lambda_t) &= \sum_{j=1}^{M} P(x_0, x_1, x_2, \ldots, x_t, q_t = s_j \mid \lambda_t) \\
&= \sum_{j=1}^{M} \alpha_t^{(t)}(j), \quad 1 \le j \le M.
\end{aligned}
\quad [2.34]
$$

Based on the last formulation, the probability that an OS $X^t = \{x_0, \ldots, x_t, x_{t+1}\}$ evolves to the DS S_j at the time instant t is given as

$$P(x_0, x_1, x_2, \ldots, x_t, x_{t+1}, q_{t+1} = S_j \mid \lambda_t) = \alpha_{t+1}^{(t)}(j) = b_j(x_{t+1}) \sum_{i=1}^{M} \alpha_t^{(t)}(i) a_{ij}^{(t)}. \quad [2.35]$$

The determination of the DS at the time instant t+1 passes by the research of the DS S_j, which maximizes formula [2.35]. This formulation is expressed as

$$\text{Degradation state } S_j \text{ of the system at } (t+1) = \underset{1 \le j \le M}{\arg\max} \left(b_j(x_{t+1}) \sum_{i=1}^{M} \alpha_t^{(t)}(i) a_{ij}^{(t)} \right). \quad [2.36]$$

The observation probability $b_j(x_{t+1})$ is unknown. We assume that there is a high probability of observing a new DS at the moment t+1. This leads to normalizing the observation probability to 1. Formula [2.36] is rewritten as

$$P(x_0, x_1, x_2, \ldots, x_t, x_{t+1}, q_{t+1} = S_j \mid \lambda_t) = \sum_{i=1}^{M} \delta_t^{(t)}(i) a_{ij}^{(t)} \quad [2.37]$$

with $\delta_0^{(t)}(i) = \pi_i^{(t)} b_i(x_0)$ and $\delta_{t+1}^{(t)}(j) = b_j(x_{t+1}) \sum_{i=1}^{M} \delta_t^{(t)}(i) a_{ij}^{(t)}$ and hence formula [2.33].

The algorithm for predicting the imminence of the next DS at t+1 is given as follows.

1. Initialization:

$$\delta_0^{(t)}(i) = \pi_i b_i(x_0).$$

2. Recurrence:

$$\text{Degradation state } S_j \text{ of the system at } (t+1) = \underset{1 \le j \le M}{\arg\max} \left(\sum_{i=1}^{M} \delta_t^{(t)}(i) a_{ij}^{(t)} \right)$$

$$\delta_{t+1}^{(t)}(j) = b_j(x_{t+1}) \sum_{i=1}^{M} \delta_t^{(t)}(i) a_{ij}^{(t)}, \quad 1 \le i, j \le M \quad \text{and} \quad t \ge 0.$$

Figure 2.16 shows the flowchart of the second prognostic method based on the evolutionary Markov model.

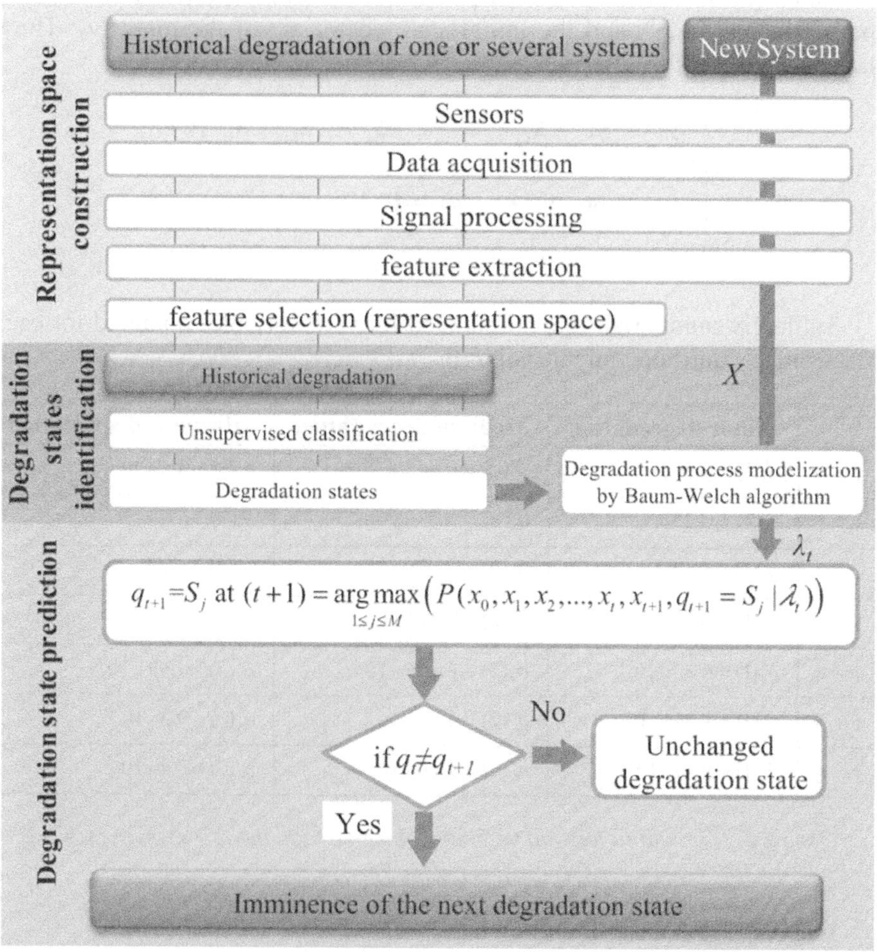

Figure 2.16. *Flowchart of the second prognostic method based on an evolutionary Markov model*

2.2.2.3.2. Back to the example

The above example is used to illustrate the applicability of the evolutionary Markov model to predict the imminence of the next DS of bearing 11. The historical degradation of bearing 1 is considered available to identify the three DSs as shown in Figure 2.6.

Based on the observation that matrix B extracted from the observation sequence X^t is measured up to time t and on the Baum re-estimation formulas [2.21] and [2.22], it is possible to obtain a re-estimation of the transition matrix $A^{(t)}$ and the initial state vector $\pi^{(t)}$ of the model λ_t. These latter are given as follows:

$$A^{(t)} = \begin{matrix} & \begin{matrix} S_1 & S_2 & S_3 \end{matrix} \\ \begin{matrix} S_1 \\ S_2 \\ S_3 \end{matrix} & \begin{pmatrix} 0.69 & 0.21 & 0.1 \\ 0 & 0.66 & 0.34 \\ 0 & 0 & 1 \end{pmatrix} \end{matrix}, \quad \pi^{(t)} = \begin{matrix} \begin{matrix} S_1 & S_2 & S_3 \end{matrix} \\ \begin{pmatrix} 0.69 & 0.21 & 0.1 \end{pmatrix} \end{matrix}.$$

Table 2.8 summarizes the normalized forward variables obtained for each DS from X^{11} until the time instant t=6.

	Minor degradation	**Medium degradation**	**Advanced degradation**
$t=0$	$\delta_0(1)=0.5325$	$\delta_0(2)=0.04$	$\delta_0(3)=4.5\times10^{-3}$
$t=1$	$\delta_1(1)=0.226$	$\delta_1(2)=0.0398$	$\delta_1(3)=6.6\times10^{-3}$
$t=2$	$\delta_2(1)=0.08$	$\delta_2(2)=0.025$	$\delta_2(3)=6.07\times10^{-3}$
$t=3$	$\delta_3(1)=0.0255$	$\delta_3(2)=0.013$	$\delta_3(3)=3.26\times10^{-3}$
$t=4$	$\delta_4(1)=6.33\times10^{-3}$	$\delta_4(2)=6.156\times10^{-3}$	$\delta_4(3)=1.99\times10^{-3}$
$t=5$	$\delta_5(1)=1.34\times10^{-3}$	$\delta_5(2)=2.66\times10^{-3}$	$\delta_5(3)=9.3\times10^{-4}$
$t=6$	$\delta_6(1)=9.51\times10^{-5}$	$\delta_6(2)=9.49\times10^{-4}$	$\delta_6(3)=8.4\times10^{-4}$

Table 2.8. *Normalized forward variables obtained from the OS X^t up to time t=6*

For t+1=7, we have the following:

$$S_j(t+1) = \arg\max\left[\sum_{i=1}^{M} \delta_t(i) a_{ij}^{(t)}\right] \Leftrightarrow S_j(t+1=7) = \arg\max\left[\sum_{i=1}^{M} \delta_6(i) a_{ij}^{(t)}\right]$$

$$\begin{aligned} \text{For } q_{t+1} = S_1 \Rightarrow \sum_{i=1}^{M} \delta_6(i) a_{i1}^{(t)} &= \delta_6(1) a_{11}^{(t)} + \delta_6(2) a_{21}^{(t)} + \delta_6(3) a_{31}^{(t)} \\ &= (9.51\times10^{-5})\times0.71 + (9.49\times10^{-4})\times0 + (8.4\times10^{-4})\times0 \\ &= 6.75\times10^{-5} \end{aligned}$$

$$\text{For } q_{t+1} = S_2 \Rightarrow \sum_{i=1}^{M} \delta_6(i) a_{i2}^{(t)} = \delta_6(1) a_{12}^{(t)} + \delta_6(2) a_{22}^{(t)} + \delta_6(3) a_{32}^{(t)}$$

$$= (9.51\times10^{-5})\times0.2 + (9.49\times10^{-4})\times0.66 + (8.4\times10^{-4})\times0$$

$$= 6.45\times10^{-4}$$

$$\text{For } q_{t+1} = S_3 \Rightarrow \sum_{i=1}^{M} \delta_6(i) a_{i3}^{(t)} = \delta_6(1) a_{13}^{(t)} + \delta_6(2) a_{23}^{(t)} + \delta_6(3) a_{33}^{(t)}$$

$$= (9.51\times10^{-5})\times0.1 + (9.49\times10^{-4})\times0.34 + (8.4\times10^{-4})\times1$$

$$= 1.17\times10^{-3}$$

$$S_j(t+1=7) = \underset{1\le j\le M}{\arg\max}\left[\sum_{i=1}^{M} \delta_6(i) a_{ij}^{(t)}\right] = \arg\max\left\{6.75\times10^{-5} \quad 6.45\times10^{-4} \quad 1.17\times10^{-3}\right\}$$

$$\rightarrow q_{t+1} = S_3.$$

Based on the calculation of the probabilities that the observation sequence X^t arrives to the DSs S_1, S_2 and S_3, it appears that the advanced DS corresponds to the high probable DS at the time instant $t + 1 = 7$.

2.2.2.4. *Hidden Markov models*

The previous prognostic methods are based on the same principle which is modeling the degradation process of a system using a single HMM. The third method is based on modeling the *M*-degradation states by *M*-hidden Markov models λ_k, $1 \le k \le M$, from the OS X^k extracted from the *M*-classes *k* (see Figure 2.17). The OS X_k, in this case, is formulated as

$$X^k = \{x_0^k, x_1^k, x_2^k, \dots, x_T^k\}, \quad 1 \le k \le M \qquad [2.38]$$

with $\{x_0^k, x_1^k, x_2^k, \dots, x_T^k\}$ being an OS extracted from the class *k*. This extraction is obtained by a random choice of $(T+1)$-observations belonging to the class *k*. The prediction of the next DS is obtained as follows:

(1) Calculating the belonging probability of the OS $\{x_0, x_1, x_2, \dots, x_t, x_{t+1}\}$ to the model λ_k, with x_{t+1} unknown.

(2) Choose among the models λ_k the one that maximizes the probability given by [2.39].

In the case of the third prediction model, the imminence of the next DS is expressed as

$$\text{Degradation state } \lambda_k \text{ at } (t+1) = \underset{1\le k\le M}{\arg\max}(P(x_0,x_1,x_2,\ldots,x_t,x_{t+1} \mid \lambda_k)). \quad [2.39]$$

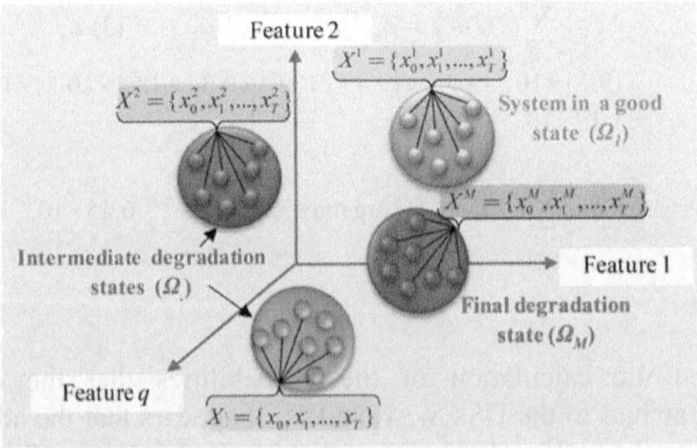

Figure 2.17. *DS modelization by an HMM*

From formulas [2.26] and [2.27], the belonging probability of the OS $\{x_0,\ldots,x_t,x_{t+1}\}$ to the model λ_k, $1 \le k \le M$, is given as

$$\begin{aligned} P(x_0,x_1,\ldots,x_t,x_{t+1} \mid \lambda_k) &= \sum_{j=1}^{M} P(x_0,x_1,x_2,\ldots,x_{t+1},q_{t+1}=S_j \mid \lambda_k) = \sum_{j=1}^{M} \alpha_{t+1}^{(k)}(j) \\ &= \sum_{j=1}^{M} b_j(x_{t+1}) \sum_{i=1}^{M} \alpha_t^{(k)}(i) a_{ij}^{(k)}, \quad 1 \le i,j \le M. \end{aligned} \quad [2.40]$$

As for the first two methods, the observation probability is unknown due to the unavailability of the observation x_{t+1}. We assume that there is a high probability of observing a new DS at the moment t+1. This leads to normalizing the observation probability to 1. A coefficient noted $C_j(x_{t+1})$ is introduced to replace $b_j(x_{t+1})$. This coefficient is between 0 and 1. A coefficient of 0 represents a low probability of observing a new DS S_j at the moment t+1 and the opposite is true for a coefficient of 1.

The prediction of the next DS is obtained by maximizing the chances of transitioning to a new stage of degradation. This is obtained by giving to the coefficient $C_j(x_{t+1})$ a value equal to 1 for the next probable DSs. Formula [2.40] is rewritten as

$$P(x_0, x_1, \ldots, x_t, x_{t+1} \mid \lambda_k) = \sum_{j=1}^{M} C_j(x_{t+1}) \sum_{i=1}^{M} \alpha_t^{(k)}(i) a_{ij}^{(k)}, \quad 1 \leq k \leq M \qquad [2.41]$$

with $C_j(x_{t+1}) = \begin{cases} 1, & \text{if } q_{t+1} \neq q_t, \\ b_j(x_t), & \text{if } q_{t+1} = q_t. \end{cases}$

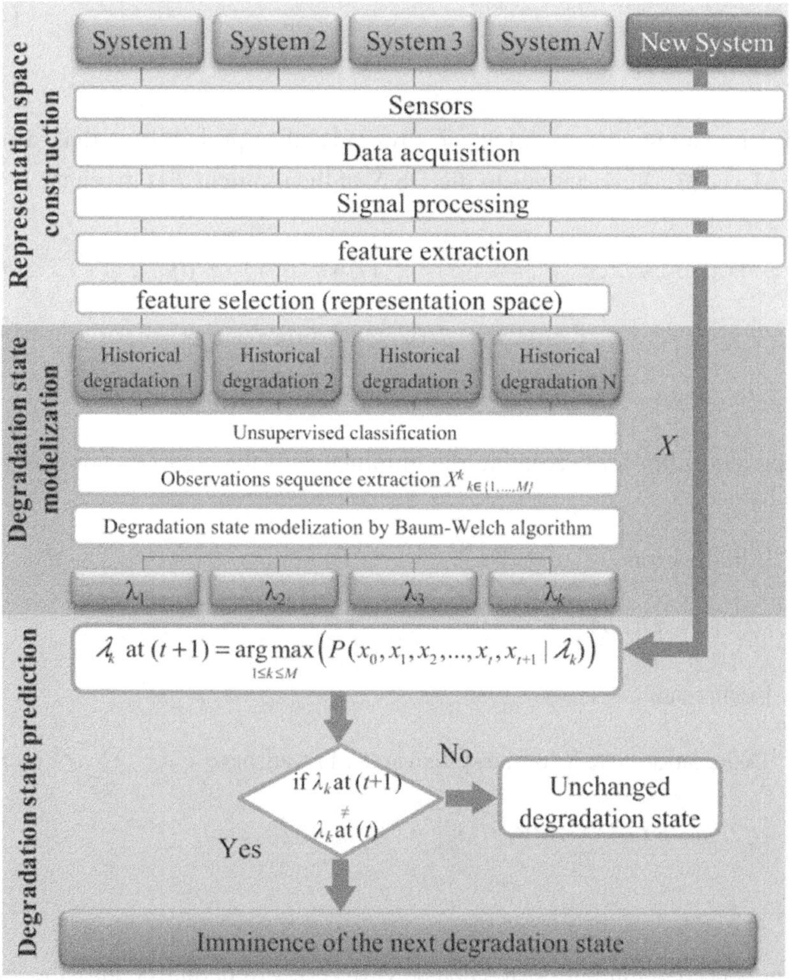

Figure 2.18. *Flowchart of the third prognostic method based on DS modelization*

Figure 2.18 shows the flowchart of the second prognostic method based on the evolutionary Markov model.

2.2.2.4.1. Proof

According to the forward and backward formulations, the probability that an OS $X = \{x_0, \ldots, x_T\}$ arrives to the DS S_j at the moment T is given as

$$\begin{aligned} P(x_0, x_1, x_2, \ldots, x_T \mid \lambda_k) &= \sum_{j=1}^{M} P(x_0, x_1, x_2, \ldots, x_T, q_t = s_j \mid \lambda_k) \\ &= \sum_{j=1}^{M} \alpha_T^{(k)}(j), \quad 1 \le j, k \le M. \end{aligned} \qquad [2.42]$$

By replacing the moment T by t+1, the probability that an OS $X = \{x_0, \ldots, x_t, x_{t+1}\}$ arrives to the DS S_j at the moment T is given as

$$P(x_0, x_1, x_2, \ldots, x_t, x_{t+1}, q_{t+1} = S_j \mid \lambda_k) = \alpha_{t+1}^{(k)}(j) = b_j(x_{t+1}) \sum_{i=1}^{M} \alpha_t^{(k)}(i) a_{ij}^{(k)}, \quad 1 \le j, k \le M \qquad [2.43]$$

with $b_j(x_{t+1}) \rightarrow C_j(x_{t+1})$ and hence the formula [2.41].

The algorithm for predicting the imminence of the next DS at t+1 is given as follows:

1. Initialization:

$$\alpha_0^{(k)}(i) = \pi_i b_i(x_0).$$

2. Recurrence:

$$\text{Degradation state } S_j \text{ of the system at } (t+1) = \underset{1 \le j \le M}{\arg\max} \left(C_j(x_{t+1}) \sum_{i=1}^{M} \alpha_t^{(k)}(i) a_{ij}^{(k)} \right)$$

$$\alpha_{t+1}^{(k)}(j) = b_j(x_{t+1}) \sum_{i=1}^{M} \alpha_t^{(k)}(i)(i) a_{ij}^{(k)}, \quad 1 \le j, k \le M, \quad t \ge 0.$$

2.2.2.4.2. Back to the example

The initial parameters of each HMM (transition matrix and the initial state vector), to start the learning by the Baum–Welch learning algorithm, are defined as

$$A^{(1,2,3)} = \begin{matrix} & S_1 & S_2 & S_3 \\ S_1 & 1/3 & 1/3 & 1/3 \\ S_2 & 0 & 1/2 & 1/2 \\ S_3 & 0 & 0 & 1 \end{matrix}, \quad \pi^{(1,2,3)} = \begin{pmatrix} \frac{1}{3} & \frac{1}{3} & \frac{1}{3} \end{pmatrix}.$$

We can observe from the transition matrix $A^{(1,2,3)}$ that the transition probabilities a_{32}, a_{31} and a_{21} are set to 0. We consider that the transition probability from a DS to another state of low level is equal to 0.

From the forward and backward formulations [2.21] and [2.22], we obtain a re-estimation of the transition matrix $A^{(k)}$ and the initial state vector $\pi^{(k)}$:

$$A^{(1)} = \begin{pmatrix} 0.99 & 0.01 & 0 \\ 0 & 0.99 & 0.01 \\ 0 & 0 & 1 \end{pmatrix}, \quad A^{(2)} = \begin{pmatrix} 0 & 0.99 & 0.01 \\ 0 & 0.99 & 0.01 \\ 0 & 0 & 1 \end{pmatrix}, \quad A^{(3)} = \begin{pmatrix} 0 & 0.01 & 0.99 \\ 0 & 0.01 & 0.99 \\ 0 & 0 & 1 \end{pmatrix},$$

$$\pi^{(1)} = \begin{pmatrix} 0.99 & 0.01 & 0 \end{pmatrix}, \quad \pi^{(2)} = \begin{pmatrix} 0 & 0.99 & 0.01 \end{pmatrix}, \quad \pi^{(3)} = \begin{pmatrix} 0 & 0.01 & 0.99 \end{pmatrix}.$$

The forward variables calculated for each degradation stage is given Table 2.9. This variable is deduced from to the OS of bearing 11 (X^{11}) evolving until the moment (t=6).

	Minor degradation	**Medium degradation**	**Advanced degradation**
k = 1	$\alpha_6^{(1)}(1) = 0$	$\alpha_6^{(1)}(2) = 6.05 \times 10^{-5}$	$\alpha_6^{(1)}(3) = 2.67 \times 10^{-6}$
k = 2	$\alpha_6^{(2)}(1) = 0$	$\alpha_6^{(2)}(2) = 4.7 \times 10^{-3}$	$\alpha_6^{(2)}(3) = 2 \times 10^{-4}$
k = 3	$\alpha_6^{(3)}(1) = 0$	$\alpha_6^{(3)}(2) = 0$	$\alpha_6^{(3)}(3) = 2.7 \times 10^{-4}$

Table 2.9. *Forward variables obtained from the OS X^{11}*

For $t + 1 = 7$, we have $C_1(x_7) = 0.1$, $C_2(x_7) = 0.47$, $C_3(x_7) = 1$ and

$$\lambda_k \; at \; (t+1) = \underset{1 \le k \le 3}{\arg\max} \left[\sum_{j=1}^{M} C_j(x_7) \sum_{i=1}^{M} \alpha_t^{(k)}(i) a_{ij}^{(k)} \right] = \underset{1 \le k \le 3}{\arg\max} \left[\sum_{j=1}^{M} C_j(x_7) \sum_{i=1}^{M} \alpha_6^{(k)}(i) a_{ij}^{(k)} \right]$$
$$= \arg\max\{2.67 \times 10^{-6} \quad 2.42 \times 10^{-3} \quad 2.7 \times 10^{-3}\} \rightarrow \lambda_k \text{ at } (t+1) = \lambda_3.$$

As can be observed from the above formula, the probability that the observation sequence X^{t1} arrives to the DS S_3 is higher than S_1 and S_2. We can conclude that the advanced DS is imminent.

Table 2.10 shows the advantages and disadvantages of each proposed method compared to their complexity and applicability.

	Advantages	**Disadvantages**
Method (1) Degradation process modelization	Complexity of $O(M^2TN)$	Needs the historical degradation of N-systems
Method (2) Online degradation process modelization	Needs the historical degradation of one system	Complexity of $O(M^2t^2N)$
Method (3) DS modelization	Complexity of $O(M^2TM)$ Needs the historical degradation of one system	–

Table 2.10. *Advantages and disadvantages of the proposed prognostic methods*

2.3. Conclusion

The prognosis of a system means the prediction of the imminence of a defect and the estimation of its remaining life (not presented in this chapter). It is in this context that we proposed three methods of prognosis. For the first method, the degradation process of each system is modeled by an HMM. For a measured sequence of observations, the solution consists of identifying among the HMMs the one that best represents this sequence. The identification of the HMM that best represents the sequence of observations makes the prediction of the imminence of the next DS it possible and thus the defect of the studied system. In the second method (online degradation process modelization), the computation of the probability that a sequence of observations arrives at the DS at the moment t+1, given the HMM modeled from this same OS, also allows us to predict the imminence of a defect. The third method predicts the imminence of a fault by modeling not the degradation process but by modeling each degradation stage.

2.4. References

[BC 05] P. Baruah and R.B. Chinnam. HMMs for diagnostics and prognostics in machining processes. *International Journal of Production Research*, 43(6):1275–1293, 2005.

[BMMB 05] R. Bensaadi, H. Mouss, M.D. Mouss, and M. Benbouzid. *Fuzzy pattern recognition based fault diagnosis*, vol. 4, pp. 347–356, January 2005.

[MOU 09] M.S. Mouchaweh. Advanced pattern recognition method for the monitoring of dynamic systems. *IFAC Proceedings Volumes*, 42(8):510–515, 2009.

[PSSR 18] T. Praveenkumar, B. Sabhrish, M. Saimurugan, and K.I. Ramachandran. Pattern recognition based on-line vibration monitoring system for fault diagnosis of automobile gearbox. *Measurement*, 114:233–242, 2018.

[QLLY 06] H. Qiu, J. Lee, J. Lin, and G. Yu. Wavelet filter-based weak signature detection method and its application on rolling element bearing prognostics. *Journal of Sound and Vibration*, 289:1066–1090, 2006.

[RAB 89] L.R. Rabiner. A tutorial on hidden Markov models and selected applications in speech recognition. In *Proceedings of the IEEE*, pp. 257–286, 1989.

[RDNWV 10] T.W. Rauber, E.M. do Nascimento, E.D. Wandekokem, and F.M. Varejao. Pattern recognition based fault diagnosis in industrial processes: review and application. In A. Herout (ed.), *Pattern Recognition*, 2010.

[SCZX 04] Q. Sun, P. Chen, D. Zhang, and F. Xi. Pattern recognition for automatic machinery fault diagnosis. *Journal of Vibration and Acoustics*, 126(6):307–316, April 2004.

3

Electrical Signatures Analysis for Condition Monitoring of Gears

3.1. Introduction

Gears are the key component of mechanical power transmission in the majority of electromechanical systems. For instance, in wind turbines, electric ships, electric railway traction systems and electric vehicles, electrical machines are linked to gears through mechanical couplings and shafts. In this respect, the early diagnosis of incipient gear faults can noticeably reduce the gear failure rate by using efficient condition monitoring techniques and hence can increase the reliability and availability of the entire electromechanical system significantly. The vibration, acoustic emission and lubricant analysis are well-known tools for the assessment of gears. The vibration analysis has been widely utilized for online condition monitoring of gears, as it reacts promptly to changes due to gear defects. The vibration types, which carry diagnostic information, can be classified into (1) transverse vibration (both axial and radial) which is mainly related to the reaction of mechanical structure to exterior perturbations and (2) torsional vibration which is directly relevant to periodic events in angular speed of rotating components such as shafts at different stages of gears [RAN 11]. Transducers such as accelerometers measure the axial or radial mechanical transverse vibration, whereas torque and speed sensors measure the torsional vibration. The transverse vibration analysis has been considered the most efficient technique for fault detection, fault diagnosis and fault prognosis in gears till date, as it requires relatively little modification to the system installation when compared with torsional vibration sensors. However, the

---

Chapter written by Shahin HEDAYATI KIA and Mohammad HOSEINTABAR MARZEBALI.

efficiency of vibration-based methods depends on the placement of sensors installed on the system. In this respect, the recent studies suggested the analysis of electrical machine signatures (EMSs), namely the stator current, the estimated rotor speed and the estimated electromagnetic torque for fault diagnosis of both electrical and mechanical faults in complex electromechanical systems. In this chapter, the major studies, which have been performed recently on this specific topic, are presented, with more emphasis on the detection of tooth localized faults (TLFs) in gears. Furthermore, an approach to establish a universal procedure that defines the response of EMSs to the torque excitation, induced by gear TLF located within the electromechanical drive train, is proposed. The proposed technique can be utilized for the feasibility study of gear TLF detection based on EMS analysis.

3.1.1. *Different kinds of gear defects*

The gear faults are classified into distributed and localized defects. Most of gear tooth irregularities such as a chipped tooth, root cracks, spalling, wear, pitting, tooth breakage and surface damage, illustrated in Figure 3.1, are considered as TLFs [JIA 06, KIA 15a]. For a healthy gear, the most important components in the vibration are related to the tooth meshing frequency and its related harmonics, which are mainly due to the modulation effect. The higher level of sideband frequency, narrowly grouped around the tooth meshing frequency, may appear in the case of distributed gear defects [RAN 11]. When a TLF occurs, the vibration spectrum includes additional sideband of the tooth meshing frequency and its harmonics, spread over a wide frequency range, spaced by the faulted gear rotation frequency [DAL 00]. Moreover, the harmonics of rotation frequencies, which are typically located at a low frequency range, may appear in the vibration spectrum [WAN 01]. It was also observed that the fault profile is directly related to both mechanical structure and torsional vibration resonances, which are provoked by the TLF when the damaged tooth is engaged [KIA 15b]. Furthermore, the feasibility study of torsional vibration analysis, based on EMSs in complex electromechanical systems, is considered an essential phase of non-invasive gear TLF detection, which was introduced for the first time by Kia *et al.* [KIA 18, KIA 19].

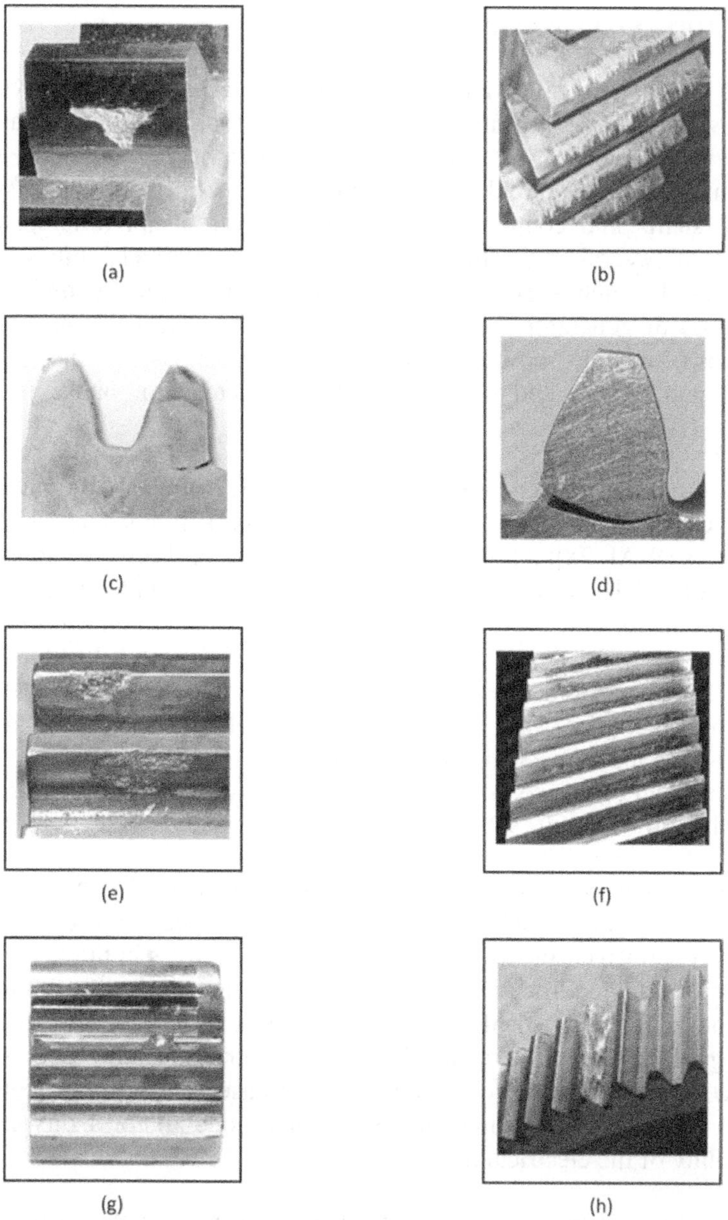

Figure 3.1. *Typical gear TLFs: (a) Macro-pitting (www.geartechnology.com); (b) scuffing [PAP 18]; (c) root cracks (www.machinerylubrication.com); (d) breakage [PAP 18]; (e) spalling [PAP 18]; (f) wear (www.machinerylubrication.com); (g) chipped tooth; and (h) surface damage. For a color version of the figures in this book, see www.iste.co.uk/soualhi/electrical1.zip*

3.1.2. *Torsional vibration study*

The torsional vibration analysis in complex electromechanical systems is rather studied for long and has been considered in the past mainly for condition monitoring of steam turbine generators in order to prevent mechanical fatigue and damage in the rotating parts of these systems, namely shafts and couplings. The rotor of large steam turbo generators consists of several rotors that are connected by relatively thin shafts and couplings. In such a system, the rotors oscillate typically owing to the interaction of generators with the power system in electrical grid transient conditions, namely power system faults, transmission line switching actions and bypass and reinsertion of series capacitor banks [CAL 79, ASP 82, GIE 12].

These electric power network transients may generate stimulus torque, which causes undamped shaft torsional natural frequencies in turbine generators [WAL 78]. Although this ancient topic has been studied in a large number of articles, a wireless sensor, which measures shaft surface strain as the main factor for the assessment of torsional vibration amplitude and frequency, was developed by the Electric Power Research Institute (EPRI) [EPR 15]. Apart from external factors linked to the electric power network transients, there are also internal issues namely mechanical load torque pulsations and mechanical imperfections of couplings in the electromechanical drive train, which yield torsional vibrations as it was originally introduced [CAR 69].

In the same way, it was observed that the gear TLF produces shaft torsional vibrations which are well identified by speed and torque sensors [FEN 13]. However, the implementation of such sensors in the system needs the installation modification and hence the usage of EMSs for torsional vibration analysis drastically simplifying the fault detection procedure. In this regard, an extensive research has been performed in recent years to study a possible torsional vibration detection based on EMSs, namely stator current and voltage measurements at the drive side and/or a flux sensor in the vicinity of the electrical machine.

Using these data, it is possible to determine or estimate other variables such as the power factor, the electromagnetic torque and the rotor mechanical speed which are also valuable for torsional vibration analysis [TRA 09, IBR 08]. Early research on this specific topic reported simply the impact of gear low-speed shaft on rotation frequencies, which may coincide with broken rotor bar frequency components in the spectrum of the stator current [THO 94, FEN 03]. Furthermore, mechanical irregularities, such as sinusoidal load torque oscillations, periodic and random dip of torque, were studied using the steady-state model of the induction machine with the aim of finding a universal technique for noninvasive mechanical anomaly detection [SAL 00].

This analysis illustrated that the crack in one of the mill rolling wheels produces a series of frequency lines in the stator current spectrum [BEL 03]. Besides, the influence of sinusoidal load torque variation on the stator current is linked to the amplitude modulation [SAL 00], whereas this concept has been associated with the stator current phase modulation [BLO 06]. Even though previous researches have studied the load variation in its general form, they can explain how EMSs can be affected by TLFs in gears.

3.2. Gear torsional vibration effects on EMSs

Based on the single-frequency torsional vibration concept, the effect of gear characteristic frequencies has been studied primarily in the stator current of induction machines [MOH 06]. The gear rotation frequencies, namely input shaft, layer shaft and output shaft frequencies in a multistage gear, have been identified in the electromagnetic torque signature. Moreover, the sideband frequencies around the supply frequency were related to the amplitude modulation of the stator current. In practice, the mesh-related torsional vibration frequencies, which are commonly damped by the system inertia, are well identified in the stator current spectrum [MOH 06, KAR 08]. In these studies, a fault detection method based on the magnitudes of some rotation- and mesh-related frequencies, sensitive to the gear tooth breakage, was proposed. However, the reasons as to why such increases or decreases in magnitudes arise in the stator current spectrum were not clearly explained. Also, it was illustrated that the vibration highlights the one-tooth breakage fault better than a tooth breakage fault with two or more teeth. This results in an important conclusion that the gear tooth fault identification may be much easier at an early stage [MOH 06].

A rigorous study was performed to develop theoretical backgrounds highlighting reasons as to why the gear mechanical characteristic frequencies would be appearing in the stator current spectrum. The idea was mainly based on the observation of the mechanical torque spectrum and the concept of multicomponent phase modulation of the stator current. This phenomenon is relevant to pinions and wheel transmission errors and gear teeth mesh stiffness variations which are the two main sources of torsional vibrations in gears [KIA 09-1, BLO 06]. It was proved analytically that, for a simple configuration in which the squirrel cage induction machine is rigidly linked to a mono-stage parallel shaft gear, the frequency components f_G related to the gear mechanical characteristics can be identified in the stator phase current spectrum at

$$f_G = \left|f_s \pm kf_p \pm mf_w \pm nf_{mesh}\right| \quad [3.1]$$

with $k = 0, 1, 2, 3, \ldots$, $m = 0, 1, 2, 3, \ldots$, $n = 0, 1, 2, 3, \ldots$ and f_s, f_p, f_w and f_{mesh} ($f_{mesh} = Z_p \times f_p = Z_w \times f_w$, where Z_p and Z_w are the number of pinion and wheel teeth) are supply, pinion rotation, wheel rotation and mesh frequencies of the parallel shaft gear, respectively. The magnitude of each frequency component in [3.1] is determined according to its related modulation index as it was discussed [KIA 09a]. An initial effort was performed using [3.1] as fault characteristic frequencies, hence their magnitude sensitivities to the gear tooth surface damage (Figure 3.1(h)) have been analyzed [KIA 12]. The results illustrate that within all eligible frequency components, $|f_s \pm f_{mesh} + f_p|$ is sensitive to the gear TLF, whereas the Welch technique is employed to reduce the effect of noise in the frequency spectrum.

A new idea is recently developed based on EMS analysis for noninvasive gear TLF detection [KIA 15]. A fault signature may appear indeed in the mechanical torque following the occurrence of TLFs in gears located in the electromechanical drive train. Owing to amplification, presented by the torsional mechanical resonance of the mechanical system, the fault profile can be identified. This specific profile produces periodic impulses in the torque signal with a defined time interval, which is related to the rotation frequency corresponding to the location of the fault (either pinion side or wheel side for a one-stage parallel shaft gear). Thus, in the stator current, the new fault-related frequency components f_{Gn} can be written as

$$f_{Gn} = \left|f_s \pm hf_{fp}\right| \quad [3.2]$$

where $h = 1, 2, 3, \ldots$ and f_{fp} is the fault profile frequency. It was shown that the most fault-sensitive frequency components in [3.2] are those that are the closest to the damped natural frequency of the drive train [KIA 15b]. Similar results have been achieved in the case of ring gear tooth cracks in a planetary gear where the permanent magnet stator current spectrum encloses rich information about the main torsional resonance frequency of the drive train [BOG 13]. Particularly, the induction machine stator current space vector instantaneous frequency (SCSVIF) was used for TLF detection in a one-stage parallel shaft gear [KIA 15b].

This last fault has been well identified through the stator current space vector instantaneous amplitude (SCSVIA) with the main assumption of stator current amplitude modulation using the time synchronous averaging technique in induction machine-based systems [OTT 13].

Thus, a universal procedure needs to be defined in order to clarify the usage of the employed technique which can evaluate the response of EMSs to torsional vibrations induced by a gear TLF located within the drive train. This leads to an initial investigation based on a model which takes into consideration the electrical machine, shafts and mechanical coupling models in the electromechanical system [KIA 18, KIA 19].

3.3. Modeling gears in complex electromechanical systems

The main aim of gear modeling in the system design stage is to study tooth and contact stress in addition to gear-mesh stiffness analysis. This approach, in complementary to analytical one, is traditionally utilized to investigate the impact of gear imperfections in the vibration signature. Moreover, using a realistic model may conduce to identify the signature of faults such as cracking and spalling in the vibration.

To analyze torsional vibrations in a representative electromechanical system, the dynamic lumped parameters model of the gear can be linked to the model of shafts and mechanical couplings, as shown in Figure 3.2. In this configuration, in which a planetary gear with a fixed-ring is located within the drive train, each mechanical component is described by its inertia, stiffness and damping coefficients.

For this purpose, m mechanical components are located on the carrier side of planetary gear, while $n–m$ ones are located at the sun gear side for the planetary gear with a fixed-ring. In this section, more emphasis is given to those studies which highlight the impact of TLFs in both parallel shaft and planetary gears in EMSs.

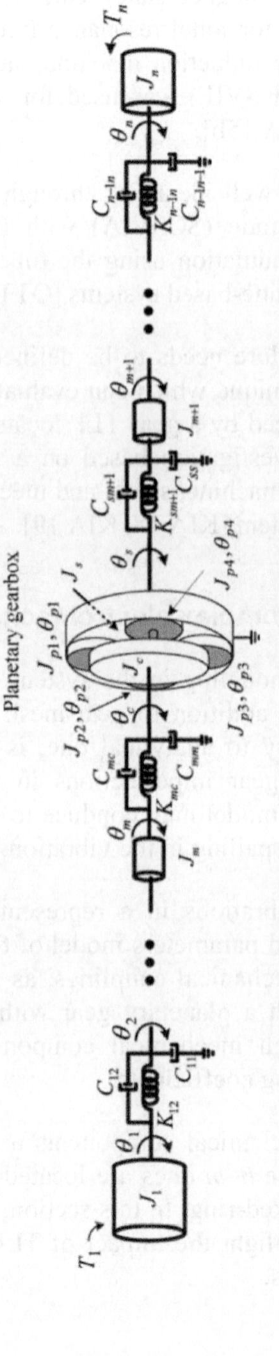

Figure 3.2. Mechanical dynamic lumped parameters model of a representative electromechanical system [MAR 16]

3.3.1. *Parallel shaft gears*

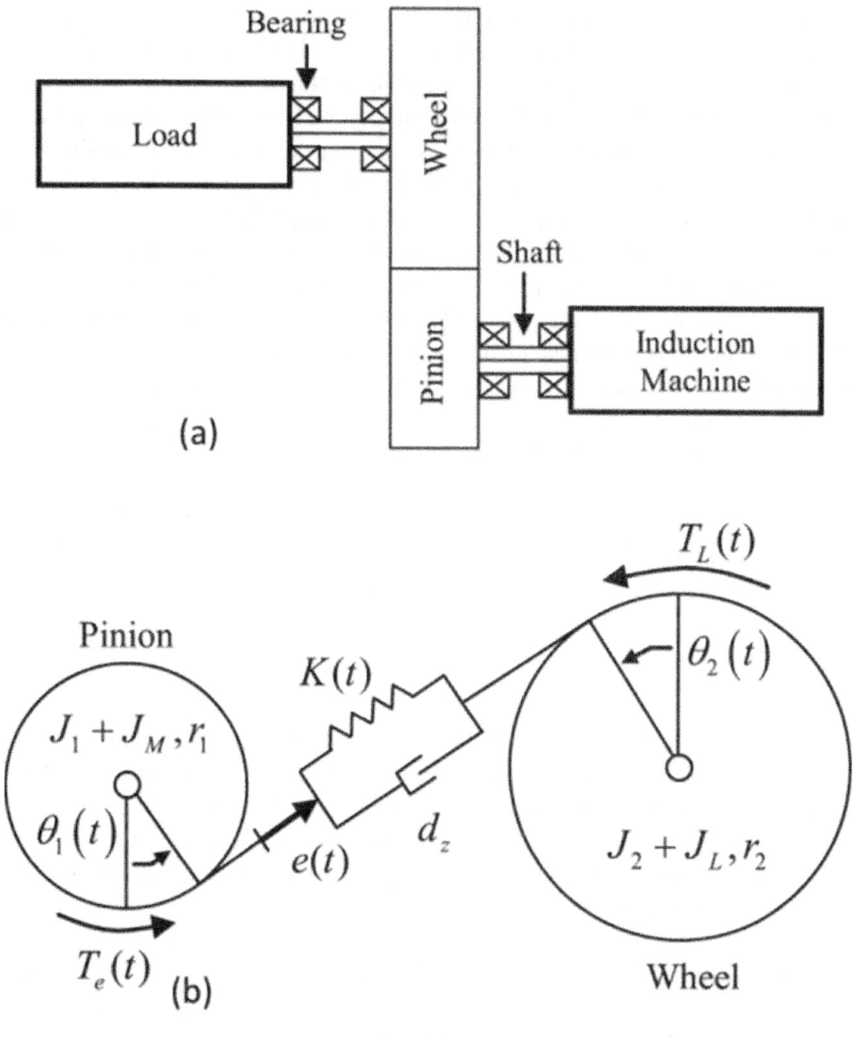

Figure 3.3. *Simplified model of the mechanical part using damping and stiffness coefficients: (a) mechanical configuration; (b) dynamic model for torsional vibration analysis [KIA 09b]*

The initial attempt studied the effect of gear torsional vibrations in EMSs exclusively with a very simple dynamic model of healthy spur gear which considers a simplified behavior of gear that needs only a minimum number of gear mechanical parameters (Figure 3.3) [KIA 09b]. This model represents an electromechanical system in which an induction machine and a mechanical load are rigidly linked to the input and output of a one-stage parallel shaft gear, respectively (Figure 3.3(a)). The main reason of stiffness time variation ($K(t)$ in Figure 3.3(b)) is the evolution of contact between the teeth of a gear. This trend depends mainly on the shape of the teeth for which the precise estimation of the stiffness variation necessitates the finite-element simulation. The stiffness variation $K(t)$ is estimated by a simplified rectangular shape which explains the phenomenon in an approximate way. $K(t)$ for each period will be at its minimum value when there is only a contact point between the pinion–wheel teeth and at its maximum value when there are two contact points [KIA 09b]. Besides, the mesh and mesh-related frequencies manifest in the stator current spectrum owing mainly to the variation of stiffness $K(t)$ in tooth contacts. The transmission error $e(t)$ in the gear model, as shown in Figure 3.3(b), is related to eccentricities of pinions and wheels, tooth profile errors and pitch errors.

These parameters ($K(t)$ and $e(t)$) produce pinion and wheel rotation sideband frequencies around the fundamental and mesh-related frequencies in the stator current spectrum described with [3.1] [KIA 09b].

3.3.2. *Planetary gears*

Planetary gears have higher efficiency, higher power-to-weight ratio, lower acoustic noise, compact and modular structure when compared with parallel shafts. It is comprised of main elements such as a sun gear, a ring gear, planet gears and a planet carrier which is a support structure for planet gears. The rotational speed of different gears in a planetary gearbox can be determined according to

$$Z'_s f'_s + Z'_p f'_p - (Z'_s + Z'_p) f'_c = 0 \qquad [3.3]$$

$$Z'_r f'_r - Z'_p f'_p - (Z'_r - Z'_p) f'_c = 0 \qquad [3.4]$$

where sun gear rotational (f'_s), planet gear rotational (f'_p) and carrier rotational (f'_c) frequencies can be computed by knowing the number of sun gear teeth (Z'_s), planet gear teeth (Z'_p) and ring gear teeth (Z'_r), respectively. The sun gear rotational speed (f'_s) can be written as a function of planet-carrier rotational speed (f'_c) for a fixed-ring gear configuration ($f'_r = 0$):

$$f'_s = ((Z'_s + Z'_r)/Z'_s)f'_c \quad [3.5]$$

In this regard, the mesh frequency (f'_{mesh}) can be obtained as [MEI 86]:

$$f'_{mesh} = ((Z'_s Z'_r)/(Z'_s + Z'_r))f'_c \quad [3.6]$$

Similar to the parallel shaft gear, the dynamic lumped parameters model of a planetary gear includes the periodically time-varying mesh stiffness which describes the evaluation of the contact between sun-planet gears' teeth and planet-ring gears' teeth in addition to the transmission error, mainly related to gears run out and damping coefficient, as it was presented [DIN 07]. This model for a one-stage planetary gear with a fixed-ring and four planets is shown in Figure 3.4.

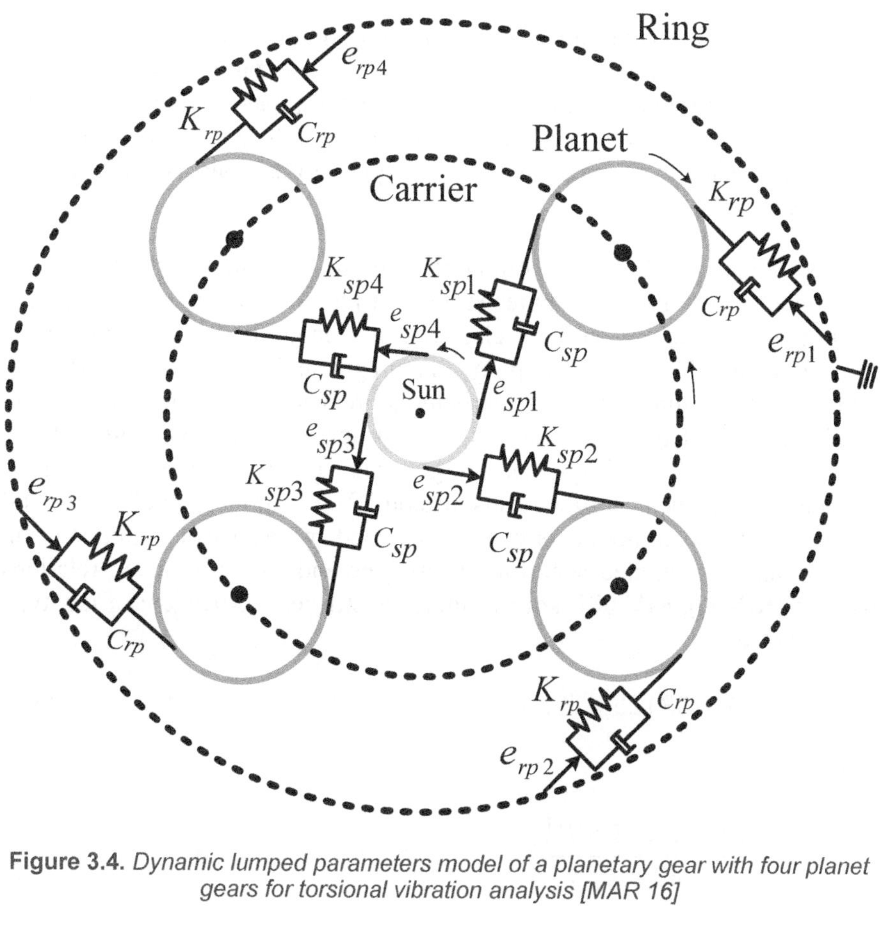

Figure 3.4. *Dynamic lumped parameters model of a planetary gear with four planet gears for torsional vibration analysis [MAR 16]*

3.3.3. *Mechanical shafts and couplings*

The torsional vibration study of any electromechanical system considers the main elements of the drive train such as shafts and mechanical couplings in its dynamic lumped parameters model [GOL 79]. It is crucial to take into account such elements in the dynamic lumped parameters model of the electromechanical system as they change the amplitude of planetary gear characteristic frequencies in the EMSs. Shafts and mechanical couplings can be considered as a simple mass–spring–damper representation in the drive train model, as shown in Figure 3.2. The stiffness coefficient (K_{sh}) of shafts is determined by [ZHA 15]:

$$K_{sh} = \frac{\pi G}{32} \times \frac{d^4{}_{sho} - d^4{}_{spI}}{L_{sh}} \quad [3.7]$$

where G is the shear modulus, d_{sho} is the outer diameter of the spline, d_{spI} is the inner diameter, and L_{sh} is the spline length. The damping coefficient of shafts (C_{sh}) can be determined as a fraction of K_{sh}, based on the loss factor of shaft materials (for steel materials, $C_{sh} = 0.002K_{sh}$) [SIL 06]. The damping coefficient changes based on the type of materials used in the shaft structure.

Mechanical couplings are commonly used to link rotating shafts to other components of the electromechanical drive in order to avoid any misalignment. They are mainly classified as beams, bellows, chains, jaws, discs, Oldham, Schmidt, clamping, diaphragm and grid kinds [LOV 00]. Among them, beam and bellow structures are widely used for transferring small mechanical torque in motion control applications [LOV 00]. Metal-based mechanical couplings, namely gear and disc types, are well adapted for high mechanical torque transmission. The damping coefficient and stiffness of these components can be achieved by means of manufacturing data or can be estimated according to its type and structure, using relations given in [RIV 86, RIV 08]. For instance, the stiffness coefficient of jaw type can be written as

$$K_c = 0.0703125 \times Z \times k_t \times {D_{ext}}^2 \quad [3.8]$$

with

$$K_t \approx \frac{\pi^2 \times E \times Z \times D_{int}^4}{7.48 \times D_{ext}^3}$$

where Z, D_{ext}, D_{int} and E are the number of spider's legs in the mechanical coupling, external and internal diameters of spider in mechanical coupling and Young modulus for the used material, respectively. The damping coefficient of mechanical coupling, based on the spider material, which connects two hubs to each other, needs to be evaluated. For instance, mutual damping coefficient (C_c) can be calculated with nitrile spider material as [MAD 93]:

$$C_c = 0.2 \times K_c \quad [3.9]$$

3.3.4. *Electrical machine model*

There are various models of electrical machines in the literature, such as classical *d–q* model [FEK 12, KIA 18, ZHA 15], internal equivalent circuit techniques [MAR 16], permeance network model [FEK 13, OST 89] and 2D finite-element (FE) method [ZHA 15], which are utilized for torsional vibration studies. Obviously, the classical *d–q* model is not sufficiently accurate to consider the effect of the stator and rotor slotting, air-gap length variations, rotor eccentricities, the variations in stator and rotor inductances with saturation, and so on, which may perturb the early-stage TLF detection in gears using EMSs analysis. In this respect, the permeance network model is used for the detection of gear tooth spalls through the stator current analysis of the induction machine [FEK 13].

The FE technique is also utilized for modeling the permanent magnet synchronous machine in order to investigate the planetary gear tooth cracking fault effect in EMSs [ZHA 15]. It was indicated that this kind of modeling considers the magnetic saturation as well as static and dynamic rotor eccentricities, thus conducing to a more realistic model of electrical machines. The permeance network and FE models, which were previously utilized for accurate modeling of electrical machines, need detailed information of its internal construction parameters which may not be commonly available. However, the parameter of the *d–q* model of squirrel cage induction machines and wound rotor induction machines (WRIMs) equivalent model in *abc* reference frame can be easily calculated with some basic circuit-based tests in order to estimate the stator- and rotor-related self and mutual inductances. Particularly, the *abc* model of WRIM has been previously developed for torsional vibration analysis (Figure 3.5) [MAR 16]. Also, any kind of asymmetry in both stator and rotor windings can be studied using this model.

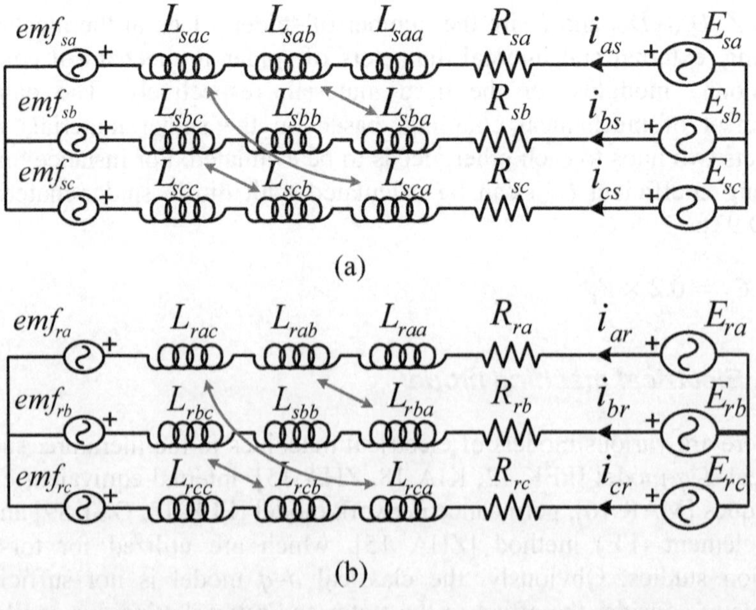

Figure 3.5. *Equivalent circuit of a WRIG: (a) stator; (b) wound rotor [MAR 16]*

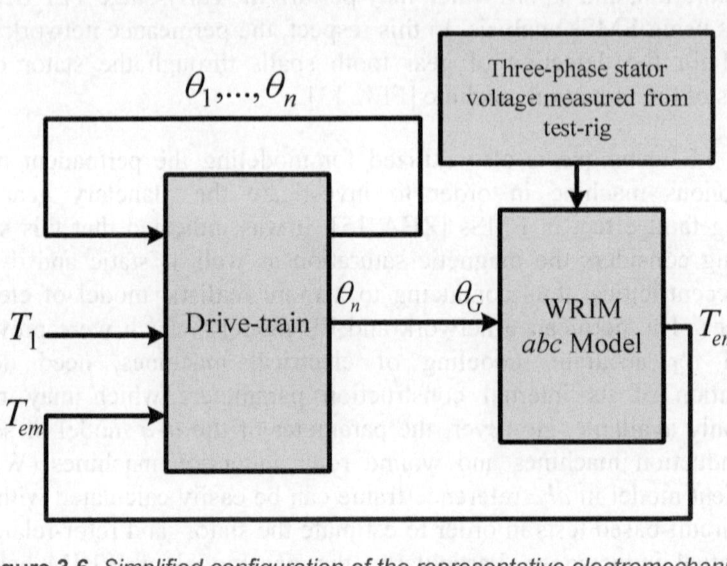

Figure 3.6. *Simplified configuration of the representative electromechanical system for numerical simulation [MAR 16]*

The global equation of the motion of complete electromechanical system, including all drive-train components and the electrical machine in matrix form, shown in Figure 3.2, can be written as [RAO 04]:

$$\mathbf{M}\ddot{\mathbf{X}}(t) + \mathbf{C}_{\mathrm{m}}\dot{\mathbf{X}}(t) + \mathbf{K}_{\mathrm{m}}(t)\mathbf{X}(t) = \mathbf{F}(t) \qquad [3.10]$$

where $\mathbf{X}(t)$, $\mathbf{F}(t)$, $\mathbf{M}$, $\mathbf{C}_{\mathrm{m}}$ and $\mathbf{K}_{\mathrm{m}}(t)$ are rotational angles of mechanical components in the drive train, mechanical torque, mass moment of inertia, stiffness and damping coefficient matrices. The block diagram of the representative electromechanical system model (Figure 3.2) is shown in Figure 3.6. The result of numerical simulations and experiments illustrates that frequency components related to the planetary gear with fixed-ring gear configuration ($f_{\mathrm{r}}' = 0$) are located in the stator current spectrum of WRIM at [MAR 16]:

$$f'_G = \left| f_S \pm kf'_s \pm mf'_p \pm nf'_{mesh} \right| \qquad [3.11]$$

3.4. Modeling gear tooth surface damage faults

3.4.1. *Parallel shaft gears*

Several investigations have been carried out based on the use of lumped parameters dynamic model of the parallel shaft gear in order to investigate the effect of different types of TLF in the EMSs. The gear TLF such as tooth pit or spall is included in the gear model as a hole with constant widths along the tooth profile. It is shown that when the defected tooth engages into mesh, the contact between the teeth can be partially lost conducing to a local reduction of mesh stiffness, as it is depicted in Figure 3.7. Such rapid fluctuations in the mesh stiffness contact may induce a wide range of frequencies in the stator current spectrum [FEK 12, FEK 13]. Furthermore, more accurate estimation of mesh stiffness can be attained by computation of the total effective Hertzian, bending, shear and axial compressive mesh stiffness for modeling the healthy and gear tooth cracking conditions [WU 08]. These preliminary numerical simulations, which reveal the feasibility of EMSs for gear TLF diagnosis, were performed without providing any experimental results [FEK 12, FEK 13].

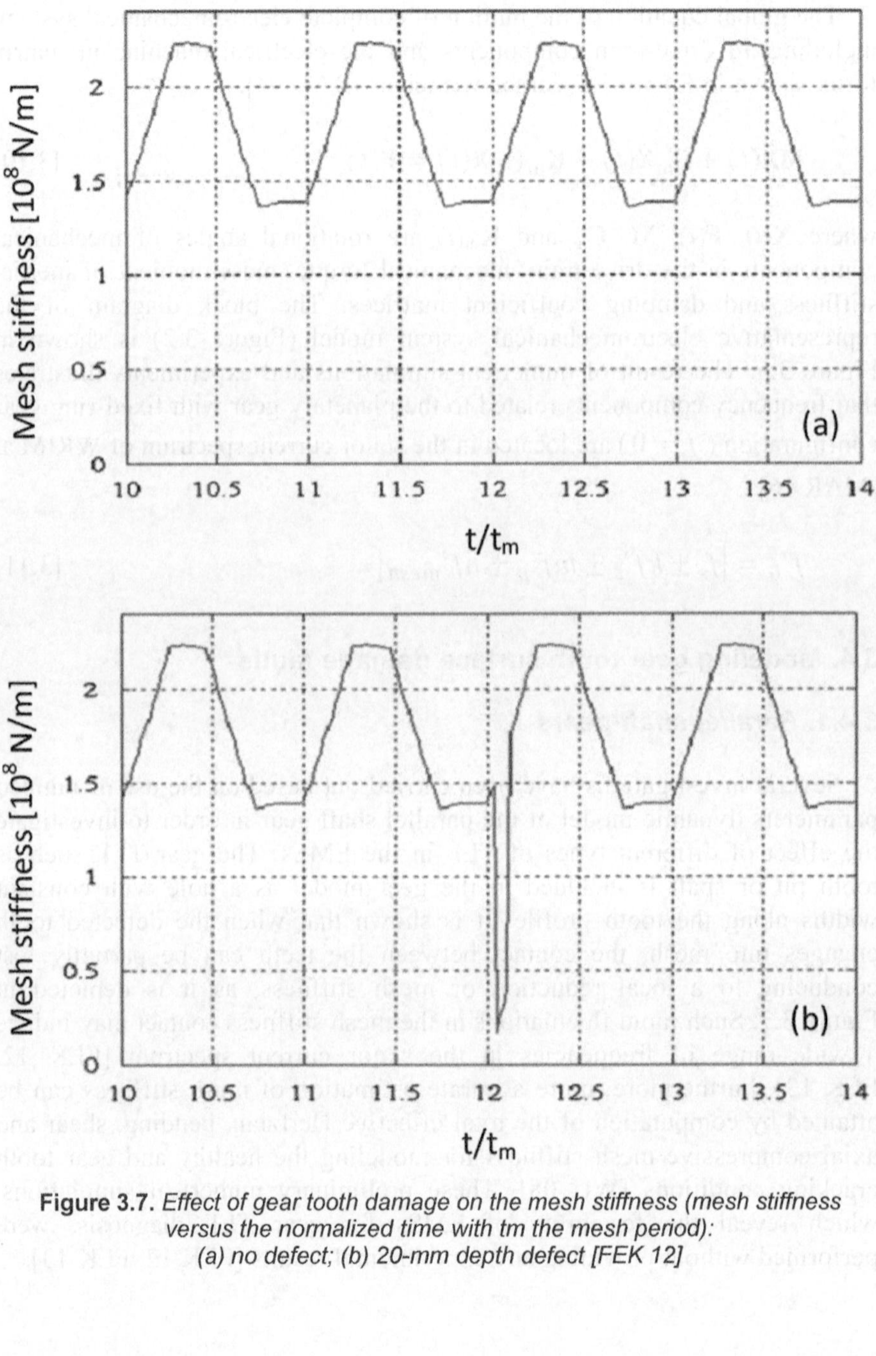

Figure 3.7. *Effect of gear tooth damage on the mesh stiffness (mesh stiffness versus the normalized time with tm the mesh period): (a) no defect; (b) 20-mm depth defect [FEK 12]*

Other relevant studies that investigate a seeded gear tooth fault in a planetary gear and its effect on the EMSs are presented [HON 13, ZHA 14]. Particularly, in [OTT 13], an initial attempt has been carried out to detect the tooth pitting fault in a multistage gear, using both numerical simulations and experiments. In this work, a low degree of freedom is used for the gear dynamic model. In the model, the tooth pitting defect is considered as a part of the static transmission error (STE) function. It was considered that the STE is equal to zero at all times, where the defective tooth was not in mesh with the drive gear (i.e., ideal tooth profile); otherwise, it can be described by a prompt variation which is depicted in Figure 3.8(a).

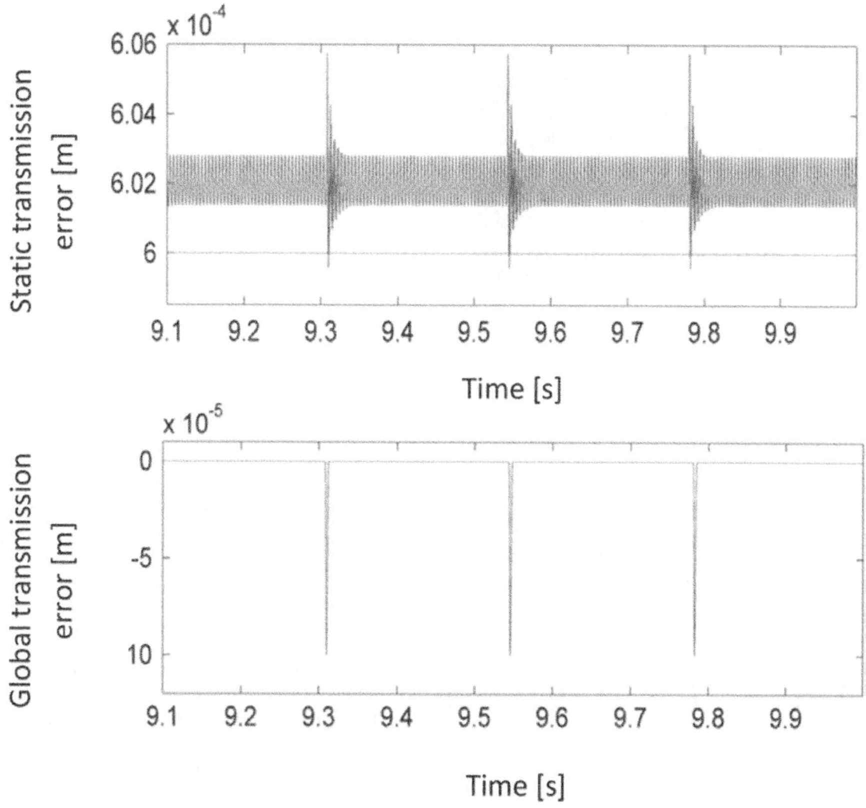

Figure 3.8. *Transmission errors in the case of gear tooth pit fault: (a) static transmission error; (b) global transmission error [OTT 13]*

These variations induce a damped oscillation in the global transmission error $e(t)$ (Figure 3.8(b)) and subsequently in the mechanical torque which produce relevant signatures in the instantaneous magnitude of the stator current space vector. In the above model, backlash nonlinearities for gear TLFs are considered by fast pulsation on both mesh and transmission error functions, as it was discussed [OTT 13].

3.4.2. Planetary gear faults

The TLF is produced in a planetary gear by removing a tooth surface of the sun gear with a depth of 0.3 mm, as shown in Figure 3.9(b) [MAR 18].

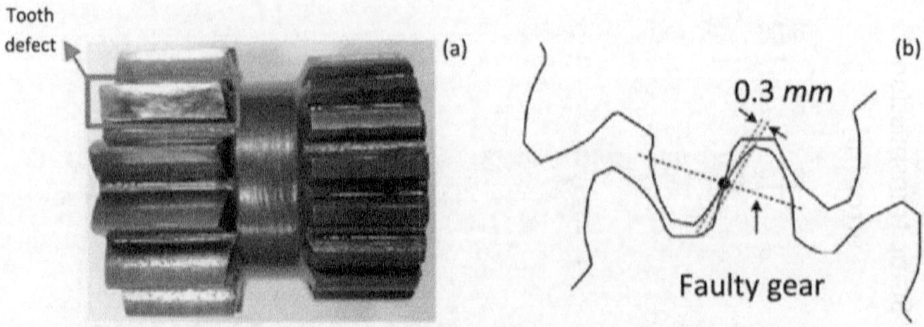

Figure 3.9. *(a) Photograph of the planetary sun gear with a defect marked. (b) A schematic of a local tooth defect approximated based on the amount of material removed from the entire face of the tooth [MAR 18]*

In this specific case, the TLF in the sun gear of the planetary gearbox (Figure 3.9(a)) can be considered in **F**(t) in [3.10] as the STE of the sun gear, where ith planet related to the sun gear fault, e_{spiF} (t), is defined by [3.12] [OTT 17]:

$$\begin{cases} \frac{\acute{E}}{2}(1-\cos(Z_s(\theta_s(t)-\theta_c(t))+\chi_i+\psi)) & \begin{array}{c}\chi_i+\psi \leq (\theta_s(t)-\theta_c(t)) mod\ 2\pi\ \& \\ (\theta_s(t)-\theta_c(t))\ mod\ 2\pi \leq \chi_i+\psi+\frac{1}{Z_s}\end{array} \\ 0 & otherwise; \end{cases} \qquad [3.12]$$

where ψ is the phase angle and $\chi_i = (2\pi(i-1))/4, i = 1,\ldots,N$, the number of planet gears with θ_s and θ_c are the rotational angular frequency of the sun gear and the planet carrier, respectively. Relation [3.12] can be customized for a planetary gear with four planet gears (Figure 3.10).

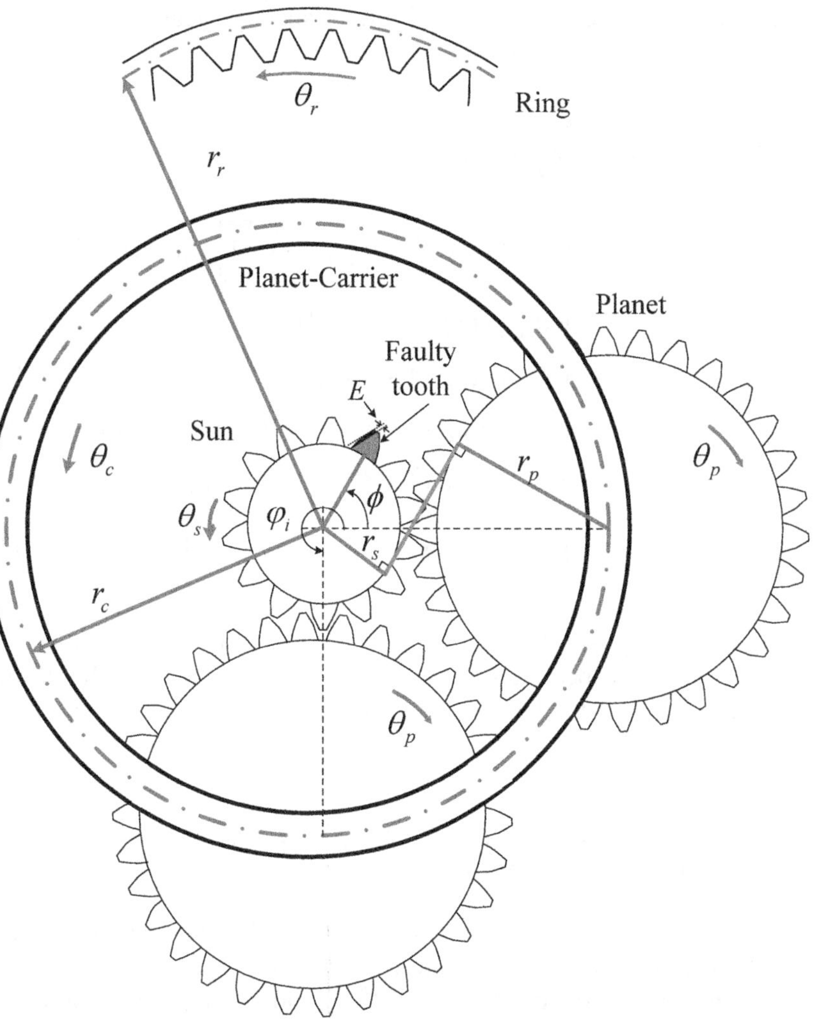

Figure 3.10. *Scheme of a planetary gearbox with four planet gears and a fixed-ring with sun TLF [MAR 18]*

The E' parameter, given in [3.12], is defined as

$$\acute{E} = \acute{E}_1 + 2\beta \tag{3.13}$$

where β is the magnitude of the sun gear inherent backlash and E_1' is the magnitude of the defect. In this respect, $\mathbf{F}(t)$ based on STE, which is defined for the TLF in the sun gear of the planetary gearbox under study (Figure 3.10), can be written as [3.14]

$$
\begin{aligned}
\mathbf{F}(t) = \Bigg[0, \dots, -\sum_{i=1}^{4} c_{rp}\acute{e}_{rpi}(t) \\
&+ c_{sp}\left(\acute{e}_{spi}(t) + \acute{e}_{spi_F}(t)\right), \sum_{i=1}^{4} c_{sp}\left(\acute{e}_{spi}(t)\right. \\
&\left. + \acute{e}_{spi_F}(t)\right), -c_{rp}\acute{e}_{rp1}(t) \\
&+ c_{sp}\left(\acute{e}_{sp1}(t) + \acute{e}_{sp1_F}(t)\right), -c_{rp}\acute{e}_{rp2}(t) \\
&+ c_{sp}\left(\acute{e}_{sp2}(t) + \acute{e}_{sp2_F}(t)\right), -c_{rp}\acute{e}_{rp3}(t) \\
&+ c_{sp}\left(\acute{e}_{sp3}(t) + \acute{e}_{sp3_F}(t)\right), -c_{rp}\acute{e}_{rp4}(t) \\
&+ c_{sp}\left(\acute{e}_{sp4}(t) + \acute{e}_{sp4_F}(t)\right), \dots, 0\Bigg]
\end{aligned}
\qquad [3.14]
$$

where c_{sp} and c_{rp} are the damping coefficients of the sun-planet and ring-planet gears, respectively. The STE of the sun gear with the *i*th planet and the ring gear with the *i*th planet are shown in [3.14] as e_{spi} and e_{rp}, respectively.

3.5. Online condition monitoring of gears in complex electromechanical systems

The numerical simulation and analytical approaches highlight, specifically at their initial stages, the characteristic frequency of gears in the EMSs as the fault indicator (see both [3.1] and [3.11]). Nevertheless, it is recently observed that the torsional resonance analysis is an interesting approach that helps identify efficiently the gear TLF in complex electromechanical systems. It is verified that the TLF may generate mechanical impacts which are recognized both in the mechanical torque and in EMSs. The effect of gear faults can be amplified in a frequency bandwidth based on the natural frequencies of the mechanical system. This approach has confirmed its efficacy for TLF detection using EMS analysis for a defined electromechanical system [KIA 15b]. A complete set of motion equations are used then to determine the natural frequencies of the drive train depicted in Figure 3.2. Eigenvalues or natural frequencies can be calculated when the effects of damping ($\mathbf{C}_m$) and applied forces ($\mathbf{F}(t)$) are ignored. Thus, expression [3.10] can be simplified as

$$\mathbf{M}\ddot{\mathbf{X}}(t) + \mathbf{K}_m(t)\mathbf{X}(t) = \mathbf{0} \quad [3.15]$$

where the displacement vector ($\mathbf{X}(t)$) with sinusoidal vibration assumption is described as

$$\{\mathbf{X}(t)\} = \{\emptyset\}e^{i\omega t} \qquad \omega t\text{: Phase angle}$$

$$\omega = 2\pi f_n \qquad f_n\text{: Natural frequencies.} \quad [3.16]$$

Substituting [3.16] to [3.15] results in

$$(-\omega^2\mathbf{M} + \mathbf{K}_m)\{\emptyset\}e^{i\omega t} = \mathbf{0} \quad [3.17]$$

since $e^{i\omega t}$ is nonzero, and [3.17] can be rearranged in the form of a general eigenvalues problem:

$$(\mathbf{K}_m - \lambda\mathbf{M})\{\emptyset\} = \mathbf{0} \quad [3.18]$$

with $\lambda = \omega^2$. In [3.18], λ represents the eigenvalues related to the natural frequencies of the mechanical system. A test bench, based on a WRIM, which has been mainly designed for wind turbine condition monitoring studies, is used for validation of the above concept (Figure 3.11(a)) [MAR 16]. This system consists of a three-phase 5.5 kW, 50 Hz, 380/190 V, 8-pole, 800 rpm and 17/19 A WRIM, a 11 kW squirrel cage induction gear-motor unit with 10.9 gear ratio and 125 rpm rated speed as a prime mover controlled by an AC/AC drive, a planetary gear with 6.57 gear ratio, a main shaft, four couplings, three bearings and a torque transducer with 50 mV/N.m sensitivity (Figure 3.11(a)). The stator of WRIM is connected to the grid through an auto-transformer. The configuration of the dynamic lumped parameters model of this electromechanical system is shown in Figure 3.11(b). More information concerning the detailed structure of the test bench can be found in [MAR 16].

3.5.1. *Mechanical torque and stator current analysis*

At the preliminary stage, the torsional natural frequencies of the drive train, presented in Figure 3.11(a), are computed using [3.18]. Moreover, the equivalent dynamic lumped parameters model is obtained through relations [3.7]–[3.9] (see Table 3.1).

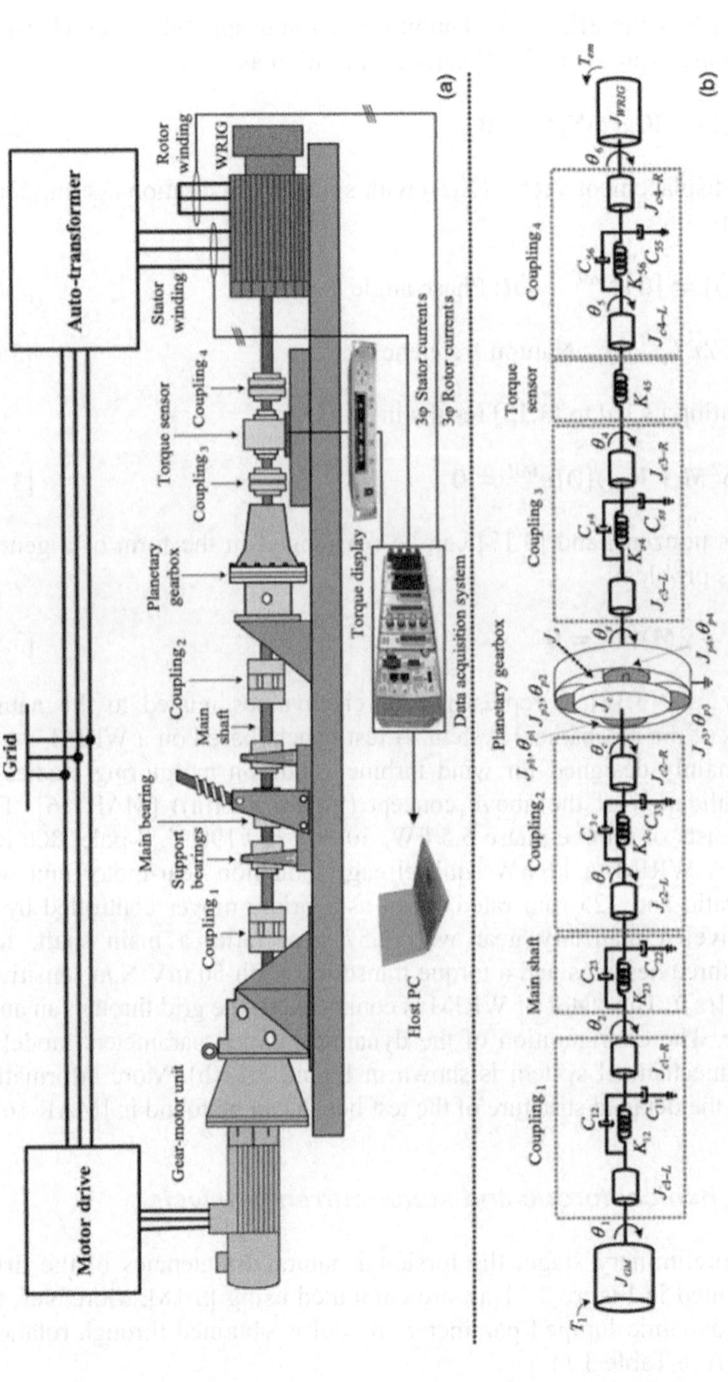

Figure 3.11. *An electromechanical system for wind turbine condition monitoring studies: (a) experimental system configuration with the data acquisition system; (b) equivalent dynamic lumped parameters model [MAR 16]*

Natural frequencies (Hz)	Mechanical component(s) in the drive train which excite(s) the natural frequencies
11	Main shaft
37	Coupling_4
73	Coupling_3
180	Torque transducer
746	Coupling_1
808	Coupling_2
8,711	Planetary gear
11,729	Planetary gear
12,081	Planetary gear

Table 3.1. *Torsional natural frequencies of the drive train*

The mechanical torque and stator currents of WRIM are collected for 180 s at 5 kHz sampling frequency at each load level. This gives 18 data windows of 10 s for both signals. Furthermore, several experiments have been performed under different load levels in order to study the behavior of the fault profile harmonic amplitudes. It is noted that the mechanical torque spectrum is normalized to the rated torque value of WRIM which is 75 N m in this case [MAR 18]. The torque transducer output, which is implemented close to the shaft of the sun gear, is analyzed. This sensor gives a good description of the TLF (Figure 3.9(a)) and leads to the observation of fault impacts in the stator current of WRIM. In such a case, the fault profile frequency f_{fp} can be obtained in the measured torque [HON 13]:

$$f_{fp} = N \times (f'_s - f'_c) = ((N \times Z'_r)(Z'_s + Z'_r))f'_s \qquad [3.19]$$

where N is the number of plant gears. The fault profile harmonics ($j \times f_{fp}$ with $j = 1, 2, 3, \ldots$) is easily observed in the spectrum of measured torque at rated load condition of the WRIM (Figure 3.12(b)). The amplitudes of the first and second harmonics of fault profile (f_{fp} and $2f_{fp}$) in the measured torque spectrum have higher magnitudes compared to the other harmonics, as this fact is clearly depicted in Figure 3.12(b). The main reason is that the frequency components f_{fp} and $2f_{fp}$ are close to the torsional natural frequencies of the drive train, that is, 37 and 73 Hz, as shown in Figure 3.12(a), respectively.

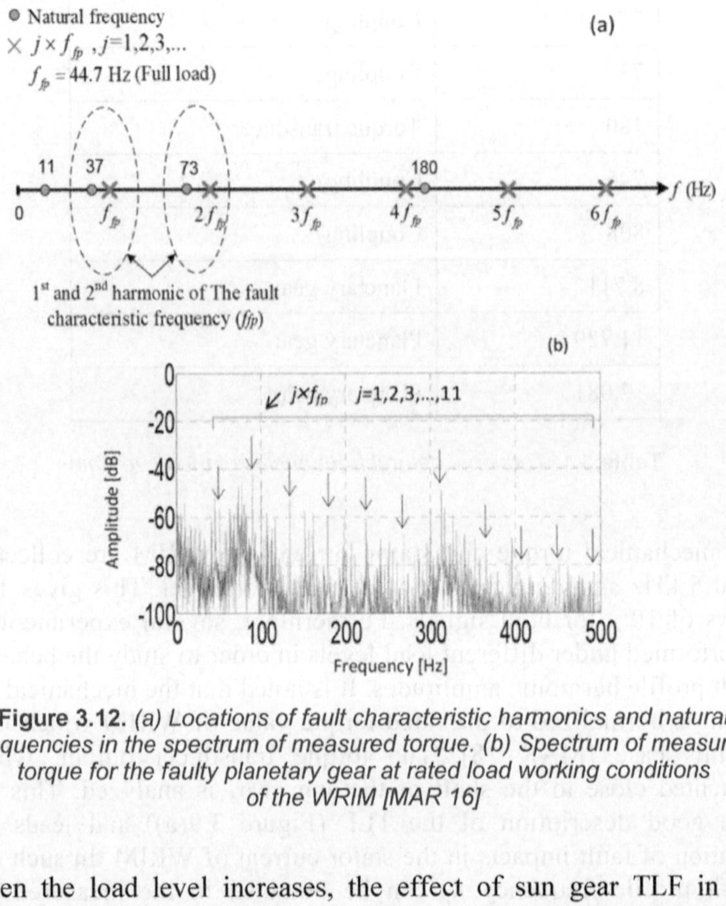

Figure 3.12. *(a) Locations of fault characteristic harmonics and natural frequencies in the spectrum of measured torque. (b) Spectrum of measured torque for the faulty planetary gear at rated load working conditions of the WRIM [MAR 16]*

When the load level increases, the effect of sun gear TLF in torque signature becomes obvious (Figure 3.13(a) and (b)). Since the amplitude of fault profile frequency (f_{fp}) becomes more obvious with the increase in the load level, it can be concluded that the detection of the fault would be simpler at higher levels of the load. Besides, the amplitudes of planetary gear mechanical characteristic frequencies (f'_s, f'_{mesh}), which have been formerly used for gear TLF diagnostic, are analyzed as well (Figure 3.13(c) and (d)) [MAR 18]. The achieved results show that the sun gear rotational frequency

(f_s') is identified in the measured mechanical torque spectrum. However, the amplitude of other frequency components such as planet gear rotational frequency and the mesh frequency are much less than the noise level, especially at low levels of the load. As it was shown, the magnitudes of these frequency components (f_s', f_{mesh}') are not affected by the sun gear TLF (Figure 3.13(c) and (d)).

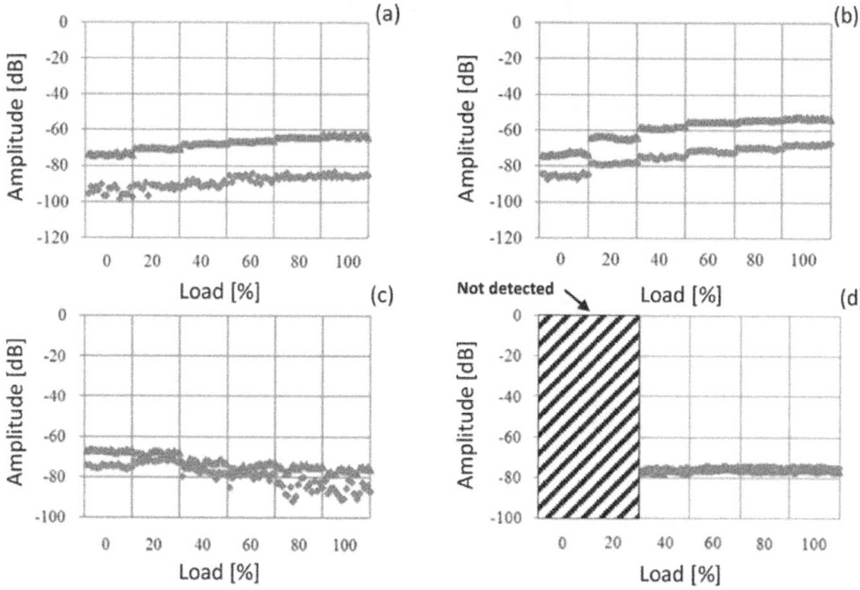

Figure 3.13. *Amplitudes of frequency components of mechanical torque related to a TLF in the planetary gearbox's sun gear: (a) f_{fp}; (b) $2f_{fp}$; (c) f'_s; (d) f'_{mesh} (◆Healthy; ▲Faulty) [MAR 18]*

The stator current spectrum at the rated load condition of the WRIM, obtained by using FFT, illustrates the presence of fault profile harmonics according to [3.2]. Spectra of all stator currents are normalized to the main frequency component of the stator current. Although $2f_{fp}$ frequency has a higher magnitude in the measured mechanical torque spectrum when compared with f_{fp}, this latter component is more sensitive to the fault in the stator current and can be considered as an appropriate fault index for noninvasive sun gear TLF detection (Figure 3.14). It should be recalled that the observed phenomenon is clearly related to the fact that the high-frequency load torque oscillation is damped in the stator current, that is, $2f_{fp}$ harmonic is damped more than f_{fp} frequency component in the stator current [KIA 09b].

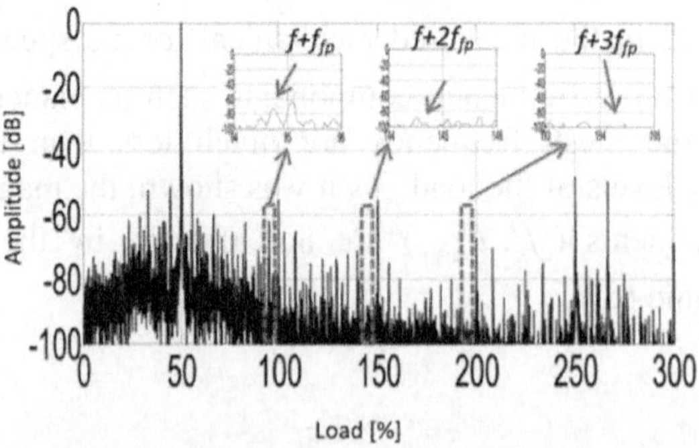

Figure 3.14. *WRIM stator current spectrum for the faulty planetary gear [MAR 18]*

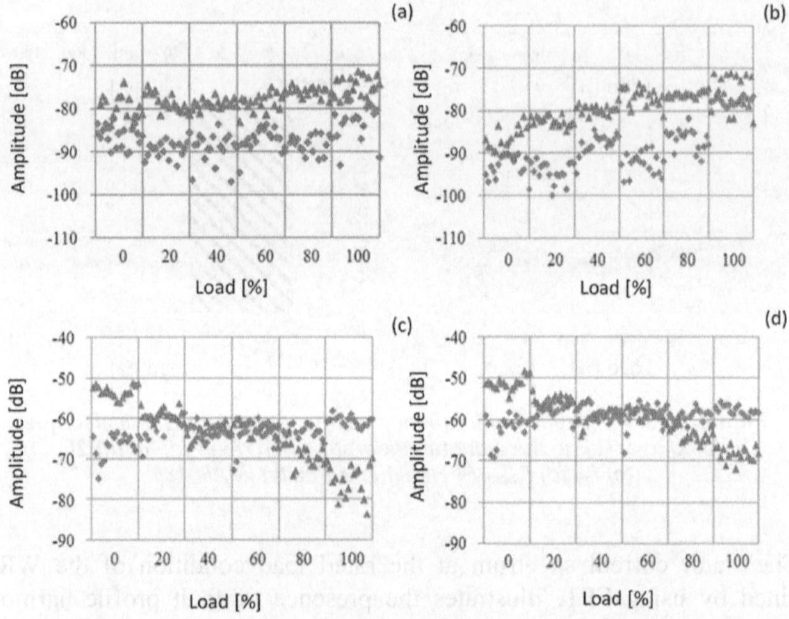

Figure 3.15. *Amplitudes of frequency components in the stator current related to a TLF in the planetary gearbox's sun gear: (a) $f_s \pm f_{fp}$; (b) $f_s + 2f_{fp}$; (c) $f_s \pm f'_s$; (d) $f_s - f'_s$ (◆Healthy; ▲Faulty)*

Moreover, among all frequency components, the magnitude of both mesh and fault characteristic frequencies are not affected by the load level in the measured torque. Similar to the torque sensor data analysis, Figure 3.15 demonstrates a meaningful increase in the amplitude of fault characteristic frequencies ($f_s \pm f_{fp}$) and ($f_s + 2f_{fp}$) when compared with the healthy planetary gear owing to the fault impacts. Therefore, the sun gear TLF can be detected in its early stage using the stator current analysis. However, the sun gear rotational frequency components ($f_s \pm f_s'$) in the stator current of WRIM have different sensitivities at high and low levels of the load (see Figure 3.15(c) and (d)). It should be noted that the amplitudes of these frequency components are not good criteria for gear TLF diagnosis since the sun gear rotational frequency can also be affected by other kinds of faults such as WRIM rotor eccentricity.

3.5.2. *Stator current space vector analysis*

The fault-related frequency components are mainly placed around the supply frequency in the stator current spectrum, while the SCSVIF and the SCSVIA give signatures similar to the mechanical toque [KIA 15]. These signals can be defined as a function of stator phase currents $I_{sA}(t)$, $I_{sB}(t)$ and $I_{sC}(t)$ in order to study the torsional perturbations induced by a gear TLF.

$$SCSVIA = \sqrt{i_D(t)^2 + i_Q(t)^2} \quad [3.20]$$

$$SCSVIF = \frac{1}{2\pi}\frac{\mathrm{d}}{\mathrm{d}t}\left(\tan^{-1}\left(\frac{i_Q(t)}{i_D(t)}\right)\right) \quad [3.21]$$

where $i_D(t)$ and $i_Q(t)$ are d-axis and q-axis stator current space vector components, respectively.

$$i_D(t) = \sqrt{\frac{2}{3}}I_{SA}(t) + \sqrt{\frac{1}{6}}I_{SB}(t) - \sqrt{\frac{1}{6}}I_{SC}(t) \quad [3.22]$$

$$i_Q(t) = \sqrt{\frac{1}{2}}I_{SB}(t) - \sqrt{\frac{1}{2}}I_{SC}(t) \quad [3.23]$$

In this regard, it is much easier to process the fault profile harmonics $j \times f_{fp}$ in both SCSVIF and SCSVIA spectra, as it was initially discussed [KIA 15b]. The performance of these signals for gear TLF detection is evaluated by using a test bench, designed based on a 250 W, 50 Hz, 400 V, star-connected, 0.77 A, 4-pole, 1,380 rpm, 24 rotor bars, three-phase squirrel-cage induction motor, which is connected to a digital controllable brake through a one-stage parallel shaft spur gear with a number of teeth at the input $N_{r1} = 25$ and at the output $N_{r2} = 75$. f_{r1} and f_{r2} represent the rotation frequencies at the input and output stages of the parallel shaft gear, respectively (Figure 3.16(a)).

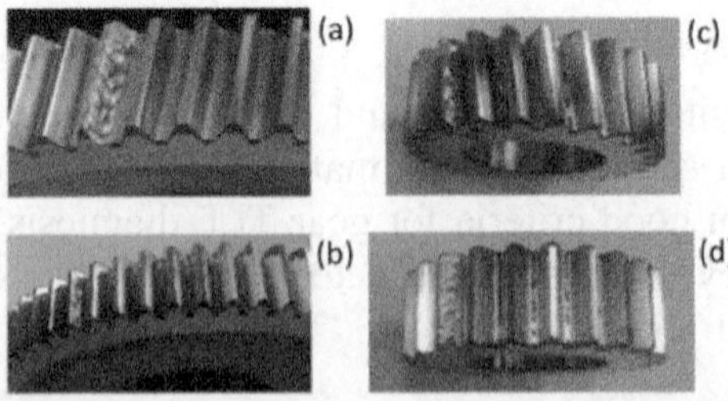

Figure 3.16. *Parallel shaft gear TLFs: (a) Helical gear wheel tooth surface damage fault. (b) Spur gear wheel tooth surface damage fault. (c) Helical gear pinion tooth surface damage fault. (d) Spur gear pinion tooth surface damage fault [KIA 16]*

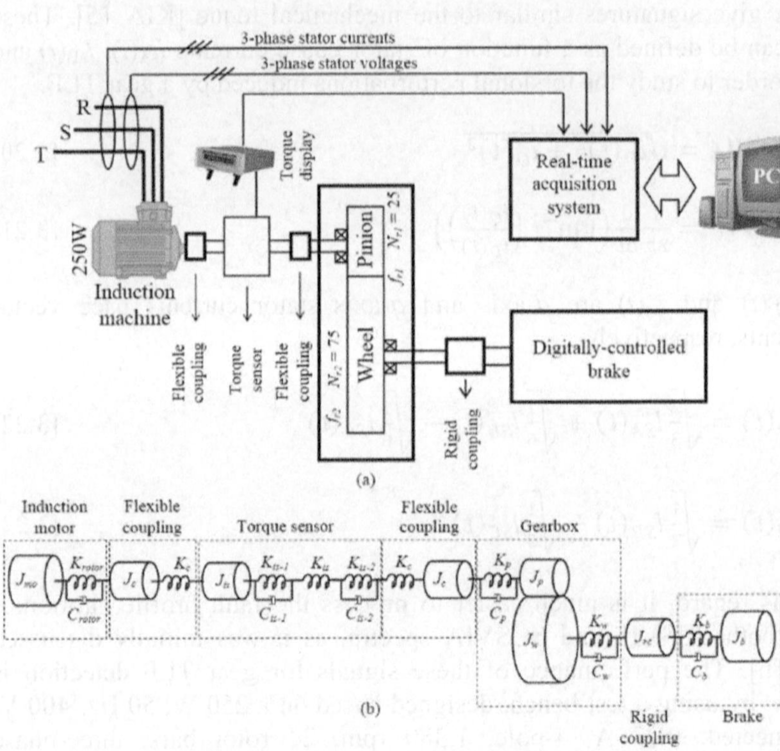

Figure 3.17. *SVSCIA and SVSCIF analyses for TLF detection in parallel shaft gears: (a) Scheme of the proposed test rig. (b) Mechanical dynamic lumped parameters model of the proposed test rig [KIA 18]*

In this test rig, the digital controllable brake system can simulate the load by keeping the rotation speed constant at the output stage of the gear. The detail of the instrumentation system can be found in [KIA 15b]. In addition, a torque sensor with 5 kHz frequency bandwidth is installed between the induction machine shaft and the input stage of the parallel shaft gear for torsional vibration measurement. Thus, the fault frequency related to SCSVIF and SCSVIA $f_{G\text{-}SV}$ can be defined as

$$f_{G-SV} = h \times f_{fp} \quad [3.24]$$

A fault detection algorithm is then proposed with the main option of fault index (FI_g) computation by means of fault profile harmonics energy evaluation in the spectrum of SCSVIF [KIA 16]. This algorithm requires the knowledge of the main natural frequency of the mechanical setup shown in Figure 3.17(a). The fault index (FI_g) is computed for detecting TLFs (Figure 3.16) in spur and helical parallel shaft gears in real time. The results are illustrated in Figures 3.18 and 3.19.

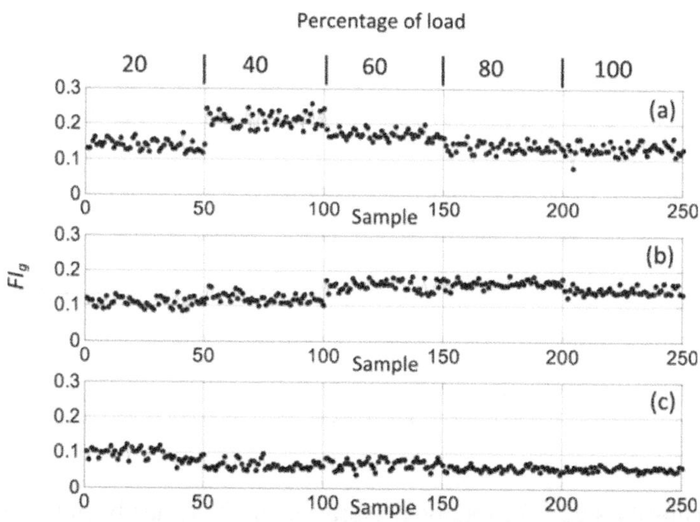

Figure 3.18. *Fault index computation (FI_g) for spur gear: (a) Pinion tooth surface damage fault. (b) Wheel tooth surface damage fault. (c) Healthy gear [KIA 16]*

In these experiments, the sampling frequency and the acquisition time have been fixed to 2 kHz and to 4 s, respectively. A first-order derivative has been implemented for the computation of FI_g through SCSVIF to achieve an acceptable error. Furthermore, the effect of the load on FI_g has been investigated by varying the load level from its minimum value (20% of the rated load) to its rated value for each set of 50 collected samples. All tests have been carried out in the steady-state condition. A primary statistical analysis has been carried out on the collected data. It has been observed that the worst case is related to the wheel TLF at the minimum load level for which the wheel faulty tooth is not completely in contact with the drive gear for both spur and helical parallel shaft gears. For the load levels greater than the minimum load value, mean values of the FI_g in faulty conditions are higher than the healthy ones for both spur and helical parallel shaft gears. The highest FI_g mean value is related to 40% of the rated load. As it is clearly depicted in both Figures 3.18 and 3.19, this value decreases once the load level increases.

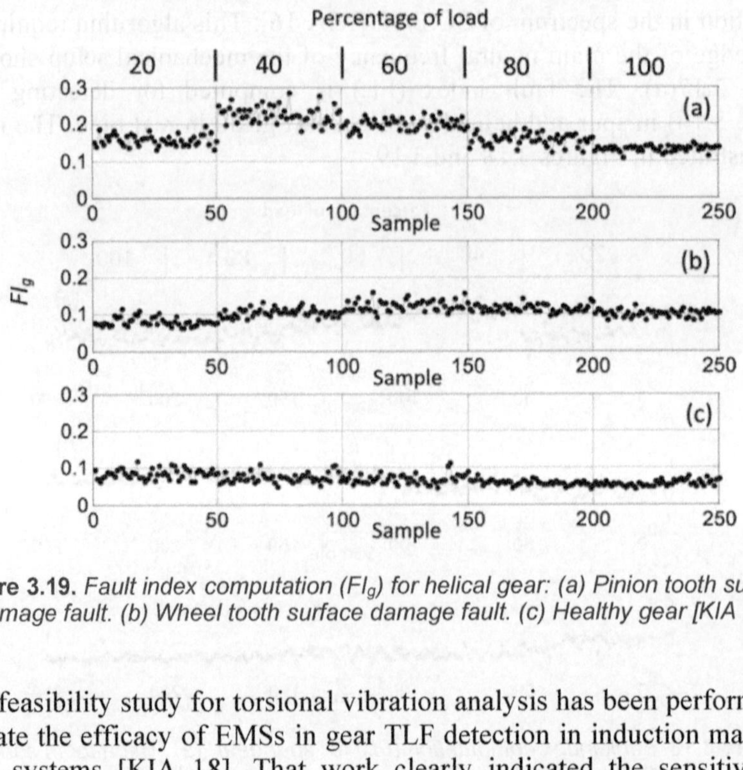

Figure 3.19. *Fault index computation (FI_g) for helical gear: (a) Pinion tooth surface damage fault. (b) Wheel tooth surface damage fault. (c) Healthy gear [KIA 16]*

A feasibility study for torsional vibration analysis has been performed to evaluate the efficacy of EMSs in gear TLF detection in induction machine-based systems [KIA 18]. That work clearly indicated the sensitivity of SCSVIF and SCSVIA to the gear TLF. In this respect, a simplified dynamic lumped parameters model of a one-stage parallel shaft spur gear is utilized. In this model, it is supposed that the stiffness of contacts between gear teeth is greater than that of gear shaft stiffness (Figure 3.17(b)). Moreover, the proposed model is validated through experiments. To evaluate the feasibility of EMSs in gear TLF detection, a torque oscillation with an amplitude of 0.1 N m (≈6% of the rated torque of induction machine under study) and a frequency between 0 and 350 Hz is applied to the output stage of the spur gear accordingly.

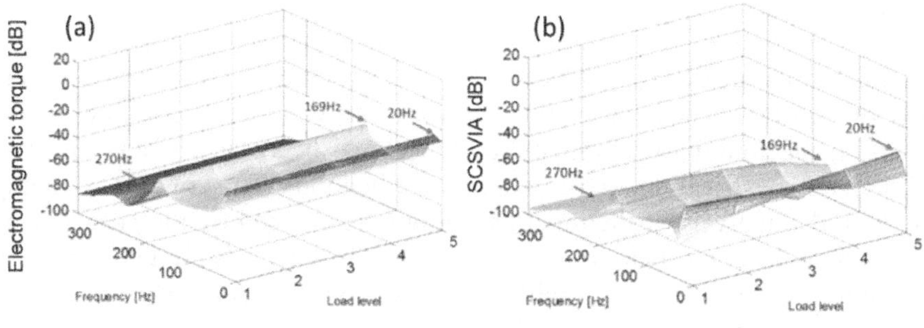

Figure 3.20. *Response of EMSs to the torque oscillation at the output stage of the gear: (a) AC part of electromagnetic torque. (b) AC part of SCSVIA [KIA 18]*

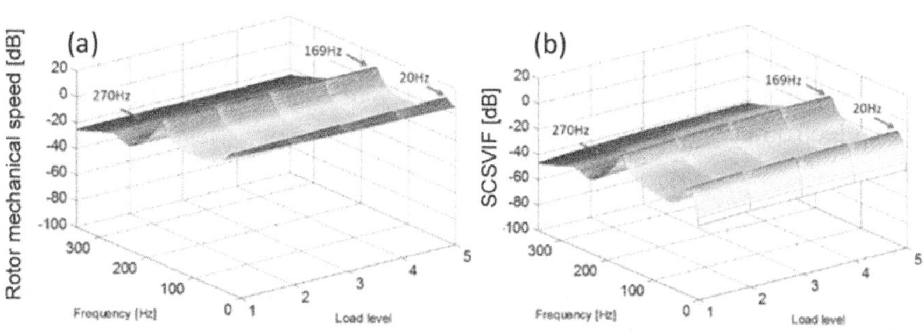

Figure 3.21. *Response of EMSs to the torque oscillation at the output stage of gear: (a) AC part of rotor mechanical speed. (b) AC part of SCSVIF [KIA 18]*

The results of numerical simulations for variables such as rotor mechanical speed [TRA 09], electromagnetic torque [KIA 09b], SCSVIA [OTT 13] and SCSVIF [KIA 15b] are illustrated in Figures 3.20 and 3.21 at five levels of the load. As it can be clearly observed in these figures, the rotor mechanical speed is the most sensitive variable to the induced torsional vibrations, but its estimation needs the knowledge of a complete set of induction machine parameters, as it was previously mentioned [TRA 09]. Therefore, as it is shown in Figure 3.21, SCSVIF is the best choice, since it is totally noninvasive and can be determined without the knowledge of induction machine internal parameters. For this drive train, the main resonance frequency is equal to 169 Hz (Figures 3.20 and 3.21). Therefore, this last phenomenon is more relevant to the stator current phase modulation, as it was suggested for this specific electromechanical system [KIA 15b].

3.6. Conclusion

This chapter presented an overview of recent research on the topic of gear TLF diagnostic, based on EMSs analysis. More attention is paid to the induction machine-based systems, as they have been largely employed in traction and electrical energy production applications. The concept is mainly relied on the analysis of the torsional vibrations induced by the gear TLF in the EMSs in a representative electromechanical system. Both analytical and modeling approaches are highlighted in this chapter. The analytical approaches link the gear TLF to the stator current phase modulation using a simplified model of the electromechanical system. However, the model-based techniques may consider all components in the drive train in order to achieve a more explicit relationship between gear TLF and EMSs. In this respect, a general approach is recently developed to study the sensitivity of EMSs to a fault located within the electromechanical system. This universal approach can be utilized for feasibility study of torsional vibration analysis in any electromechanical system.

3.7. References

[ASP 82] Aspragathos N., Dimarogonas A. D., "Fatigue damage of turbine generator shafts due to fast reclosing," *IEE Proceedings C - Generation, Transmission and Distribution*, vol. 129, no.1, pp. 1–9, Jan. 1982.

[BEL 03] Bellini A. et al., "Mechanical failures detection by means of induction machine current analysis: a case history," in *Proc. of SDEMPED'2003*, Atlanta (USA), pp. 322–326, 24–26 Aug. 2003.

[BLO 06] Blödt M., Chabert M., Regnier J., Faucher J., "Mechanical load fault detection in induction motors by stator current time-frequency analysis," *IEEE Trans. Ind. Appl.*, vol. 42, no. 6, pp. 1454–1463, Nov./Dec. 2006.

[BOG 13] Bogiatzidis I., Safacas A., Mitronikas E., "Detection of backlash phenomena appearing in a single cement kiln drive using the current and the electromagnetic torque signature," *IEEE Trans. Ind. Electron.*, vol. 60, no. 8, pp. 3441–3453, Aug. 2013.

[CAL 79] Calvaer A., Ribbens-Pavella M., Sabatel J., "Interactions between torsional vibrations of large turboalternators and electric networks," in *IEEE Conference Proc. of Power Industry Computer Applications Conference* (PICA-79), pp. 20–28, May 1979.

[CAR 69] Carter W. C., "Mechanical factors affecting electrical drive performance," *IEEE Transactions on Industry and General Applications*, no. 3, pp. 282–290, May 1969.

[DAL 00] Dalpiaz G., Rivola A., Rubini R., "Effectiveness and sensitivity of vibration processing techniques for local fault detection in gears," *Mech. Syst. Signal Process.*, vol. 14, no. 3, pp. 387–412, May 2000.

[DIN 07] Ding H., Dynamic wear model for gear system, PhD dissertation, Ohio State University, USA, 2007.

[EPR 15] Strain-based turbine generator torsional vibration monitoring system– Phase 3: Prototype field application, Technical report, Electrical Power Research Institute (EPRI), Sep. 29, 2015.

[FAI 14] Faiz J., Marzebali M.H., "Modeling and fault diagnosis of wind turbine generators: A review", in *Third International Congress on Environment and Sustainable Development,* MACDES, Cuba, 2014.

[FEK 12] Feki N., Clerc G., Velex P., "An integrated electro-mechanical model of motor-gear units-Applications to tooth fault detection by electric measurements," *Mech. Syst. Signal Process.*, vol. 29, pp. 377–390, May 2012.

[FEK 13] Feki N., Clerc G., Velex Ph., "Gear and motor fault modeling and detection based on motor current analysis", *Electric Power System Research,* vol. 95, pp. 28–37, Feb. 2013.

[FEN 13] Feng Z., Zuo M.J., "Fault diagnosis of planetary gearboxes via torsional vibration signal analysis", *Mechanical Systems and Signal Processing*, vol. 36, no. 2, pp. 401–421, Apr. 2013.

[FEN 03] Fenger M., Lyod B.A., Thomson W.T., "Development of a tool to detect faults in induction motors via current signature analysis", *Cement Industry Technical Conference, IEEE IAS/PCA* 2003, pp. 37–46, 2003.

[GIE 12] Giesecke H.D., "Steam turbine-generator torsional response due to interaction with the electrical grid" *IEEE Power and Energy Society General Meeting*, pp. 1–6, Jul. 2012.

[GIR 13] Girsang I.P., Dhupia J.S., Muljadi E., Singh M., Jonkman J., "Modeling and control to mitigate resonant load in variable-speed wind turbine drivetrain", *IEEE Journal of Emerging and Selected Topics in Power Electronics*, vol. 1, no. 4, pp. 277–286, Dec. 2013.

[GIR 14] Girsang I.P., Dhupia J.S., Muljadi E., Singh M., Pao L.Y., "Gearbox and drive train models to study of dynamic effects of modern wind turbines", *IEEE Transactions on Industry Applications*, vol. 50, no. 6, pp. 3777–3786, 2014.

[GOL 79] Golberg S., Schmus W.R., "Sub-synchronous resonance and torsional stresses in turbine-generator shafts", *IEEE Transactions on Power Apparatus and Systems*, vol. PAS-98, no. 4, pp. 1233–1237, Jul. 1979.

[HON 13] Hong L., Dhupia J.S., "A time-domain fault detection method based on an electrical machine stator current measurement for planetary gear-sets", in *Proc. of IEEE/ASME International Conference on Advanced Intelligent Mechatronics (AIM)*, Wollongong (Australia), pp. 1631–1636, 9–12 Jul. 2013.

[HUH 12] Huh K.-K., Lorenz R.D., Nagel N.J., "Gear fault diagnostics integrated in the motion servo drive for electromechanical actuators", *IEEE Transactions on Industry Applications*, vol. 48, no. 1, pp. 142–150, July/Aug. 2012.

[IBR 08] Ibrahim A., El-Badaoui M., Guillet F., Bonnardot F., "A new bearing fault detection method in induction machines based on instantaneous power factor", *IEEE Transactions on Industrial Electronics*, vol. 55, no. 12, pp. 4252–4259, Dec. 2008.

[JIA 06] Jia S., Howard I., "Comparison of localised spalling and crack damage from dynamic modelling of spur gear vibrations", *Mechanical Systems and Signal Processing*, vol. 20, no. 2, pp. 332–349, Feb. 2006.

[KAR 08] Kar C., Mohanty A.R., "Vibration and current transient monitoring for gearbox fault detection using multiresolution Fourier transform", *Journal of Sound and Vibration*, vol. 311, no. 2, pp. 109–132, March 2008.

[KIA 09a] Kia S.H., Henao H., Capolino G.-A., "Analytical and experimental study of gearbox mechanical effect on the induction machine stator current signature", *IEEE Transactions on Industry Applications*, vol. 45, no. 4, pp. 1405–1415, July/Aug. 2009.

[KIA 09b] Kia S.H., Henao H., Capolino G.-A., "Torsional vibration effects on induction machine current and torque signatures in gearbox-based electromechanical system", *IEEE Transactions on Industrial Electronics*, vol. 56, no. 11, pp. 4689–4699, Nov. 2009.

[KIA 12] Kia S.H., Henao H., Capolino G.-A., "A comparative study of acoustic, vibration and stator current signatures for gear tooth fault diagnosis", in *Proc. of ICEM'2012,* Marseille, France, pp. 1512–1517, 2–5 Sept. 2012.

[KIA 15a] Kia S.H., Henao H., Capolino G.-A., "Trends in gear fault detection using electrical signature analysis in induction machine-based systems," in *Proc. of Workshop on Electrical Machines Design Control and Diagnosis* (WEMDCD'2015), invited paper, Turin (Italy), 26-27 Mar., pp. 297–303, 2015.

[KIA 15b] Kia S.H., Henao H., Capolino G.-A., "Gear tooth surface damage fault detection using induction machine stator current space vector analysis", *IEEE Transactions on Industrial Electronics*, vol. 62, no. 3, pp. 1866–2001, Mar. 2015.

[KIA 16] Kia S.H., Henao H., Capolino G.-A., "Fault index statistical study for gear fault detection using stator current space vector analysis", *IEEE Transactions on Industry Applications*, vol. 52, no. 6, pp. 4781–4788, Aug. 2016.

[KIA 18] Kia S.H., Razavi-Far R., Saif M., "Torsional vibration identification using electrical signatures analysis in induction machine-based systems", in *Proc. of IEEE International Midwest Symposium on Circuits and Systems (MWSCAS)*, Windsor (Canada), pp. 813–816, 5–8 Aug. 2018.

[KIA 19] Kia S.H., El Hajjaji A., Marzebali M. H., "Planetary gear tooth fault detection using stator current space vector analysis in induction machine-based systems," in *Proc. of 23rd International Conference on Mechatronics Technology* (ICMT'2019), Salerno (Italy), 23–26 Oct. 2019.

[LOV 00] Lovejoy I., *The coupling handbook*, Lovejoy Inc., Downers Grove, IL, USA, 2000.

[MAD 93] Madigosky W.M., Fiorito, R.B, "Dynamic hardness tester and cure meter", in *Proc. Int. Conf. Ultrason.*, Vienna, Austria, pp. 446–450, 6–8 Jul. 1993.

[MAR 18] Marzebali M.H., Faiz J., Capolino G.A., Kia S.H., Henao H., "Gear fault detection based on stator current and mechanical torque signatures of wound rotor induction generator", *IEEE Transactions on Energy Conversion*, vol. 33, no. 3, pp. 1072–1085, 2018.

[MAR 16] Marzebali M.H., Kia S.H., Henao H., Capolino G.A., Faiz J., "Planetary gearbox torsional vibration effects on wound-rotor induction generator electrical signatures", *IEEE Transactions on Industry Applications*, vol. 52, no. 6, pp. 4770–4780, Nov.-Dec. 2016.

[MAC 45] McCann G.D., Warren C.E., Criner H.E., "Determination of transient shaft torques in turbine generators by means of the electrical; mechanical analogy", *Electrical Engineering*, vol. 64, no. 2, pp. 51–55, Feb. 1945.

[MEI 86] Meirovitch L., *Elements of vibration analysis*, McGraw-Hill, New York, 1986.

[MOH 06] Mohanty A.R., Kar C., "Fault detection in a multistage gearbox by demodulation of motor current waveform", *IEEE Transactions on Industrial Electronics*, vol. 53, no. 4, pp. 1285–1297, June 2006.

[OST 89] Ostovic V., *Dynamic of saturated electric machine*, Springer, New York, USA, 1989.

[OTT 13] Ottewill J.R., Orkisz M., "Condition monitoring of gearboxes using synchronously averaged electric motor signals", *Mechanical Systems and Signal Processing*, vol. 38, no. 2, pp. 482–498, Jul. 2013.

[OTT 17] Ottewill J.R., Ruszczyk A., Broda D., "Monitoring tooth profile faults in epicyclic gearboxes using synchronously averaged motor currents: Mathematical modeling and experimental validation", *Mechanical Systems and Signal Processing*, vol. 84, pp.78–99, Feb. 2017.

[PAP 18] Papatzimos A.K., Dawood T., Thies P.R., "Data insights from an offshore wind turbine gearbox replacement", *15th Deep Sea Offshore Wind R&D Conference*, Trondheim, Norway, 18 Jan. 2018.

[RAN 11] Randall R.B., *Vibration-based condition monitoring*, John-Wiley Press, Chichester, UK, 2011.

[RAO 04] Rao S.S., *Mechanical vibrations*, 4th ed., Pearson, Upper Saddle River, NJ, USA, 2004.

[RIV 86] Rivin, E.I., "Design and application criteria for connecting couplings", *Journal of Mechanisms, Transmissions, and Automation in Design*, vol. 108, no. 1, pp. 96–104, Mar.1986.

[RIV 08] Rivin, E.I., "Selection and performance criteria for power transmission coupling", Part II. Available at: www.powertransmission.com, 2008.

[SAL 00] Salles G., Filippetti F., Tassoni C., Grellet G., Franceschini G., "Monitoring of induction motor load by neural network techniques", *IEEE Transactions on Power Electronics*, vol. 15, no. 4, pp. 762–768, Jul. 2000.

[SAM 05] Samuel P.D., Pines D.J., "A review of vibration-based techniques for helicopter transmission diagnostics", *Journal of Sound and Vibration*, vol. 282, no. 2, pp. 475–508, Apr. 2005.

[SHA 09] Shah D., Nandi S., Neti P., "Stator inter turn fault detection of doubly fed induction generators using rotor current and search coil voltage signature analysis", *IEEE Transactions on Industry Applications*, vol. 45, no. 5, pp. 1831–1842, Sep./Oct. 2009.

[SIL 06] De Silva, C.W., *Vibration: Fundamentals and practice*, CRC Press, Boca Raton, FL, 2006.

[THO 94] Thomson W.T., "On-line current monitoring to detect electrical and mechanical faults in three-phase induction motor drives", in *Proc. Int. Conf. Life Manage. Power Plants*, pp. 66–73, Dec. 1994.

[TOL 12] Toliyat H.A., Nandi S., Choi S., Meshgin-Kelk H., *Electric machines: Modeling, condition monitoring, and fault diagnosis*, CRC Press, Boca Raton, FL, 2012.

[TRA 09] Trajin B., Chabert M., Regnier J., Faucher J., "Hilbert versus Concordia transform for three-phase machine stator current time frequency monitoring", *Mechanical Systems and Signal Processing*, vol. 23, pp. 2648–2657, 2009.

[WAC 93] Wachel J.C., Szenasi F.R., "Analysis of torsional vibration in rotating machinery", in *Proc. of the 22th Turbomachinery Symposium*, Texas, USA, pp. 127–151, 13–16 Sep. 1993.

[WAL 78] Walker D.N., Bowler C.E.J., Baker D.H., "Torsional dynamics of closely coupled turbine-generators", *IEEE Transactions on Power Apparatus and Systems*, vol. PAS-97, no. 4, pp. 1458–1466, Jul. 1978.

[WAN 01] Wang W., "Early detection of gear tooth cracking using the resonance demodulation technique", *Mechanical Systems and Signal Processing* vol. 15, no. 5, pp. 887–903, 2001.

[WU 08] Wu S., Zuo M.J., Parey A., "Simulation of spur gear dynamics and estimation of fault growth", *Mechanical Systems and Signal Processing*, vol. 317, nos 3–5, pp. 608–624, Nov. 2008.

[ZHA 14] Zhang J., Dhupia J.S., Gajanayake C.J., "Model based current analysis of electrical machines to detect faults in planetary gearboxes," in *Proc. of IEEE/ASME International Conference on Advanced Intelligent Mechatronics* (AIM), Besançon, France, pp. 1616–1621, 8–11 Jul. 2014.

[ZHA 15] Zhang J., Dhupia J.S., Gajanayake C.J., "Stator current analysis from electrical machines using resonance residual technique to detect faults in planetary gearboxes", *IEEE Transactions on Industrial Electronics*, vol. 62, no. 9, pp. 5709–5721, Sept. 2015.

4

Modal Decomposition for Bearing Fault Detection

4.1. Introduction

Electrical machines are inevitable in industrial and domestic applications, for producing mechanical power or transforming it into electrical power. So, it is to be expected that electrical machines are related to huge financial variables as well as safety and reliability. Although they are robust, they still subject to faults and downtime, thus affecting their reliability performances. On the basis of defects and the type of electrical machines, faults are classified into the following three categories:

1) Stator-related faults: these include electrical failures affecting the stator winding such as short circuits, inter-turn short circuits and open circuits (Cardoso 2018).

2) Rotor-related faults: these include electrical failures affecting the rotor winding, commutator/slip ring/brush failures for all rotor-wounded machines, broken rotor bars and end rings for squirrel cage machines and permanent magnet demagnetization or cracks for permanent magnet motors.

3) Mechanical-related faults: these include bearing failures, rotor eccentricity and shaft misalignment.

Chapter written by Yassine AMIRAT, Elhoussin ELBOUCHIKHI, Claude DELPHA, Mohamed BENBOUZID and Demba DIALLO.

Figure 4.1 depicts the distribution of machine failures based on the voltage supply level of electrical machines (Cardoso 2018). Irrespective of the category of the faults, the safety and reliability of electrical machines are related directly to these faults, hence affecting the operation and maintenance costs. So, new challenges arise particularly with regard to maintenance. In this context, cost-effective, predictive and proactive maintenance are more important. Condition monitoring systems (CMS) provide an early indication of component incipient failures, allowing the operator to plan system repairs prior to complete failure. Hence, CMS is an important tool for enhancing up-time and maximizing productivity, when cost-effective availability targets must be reached.

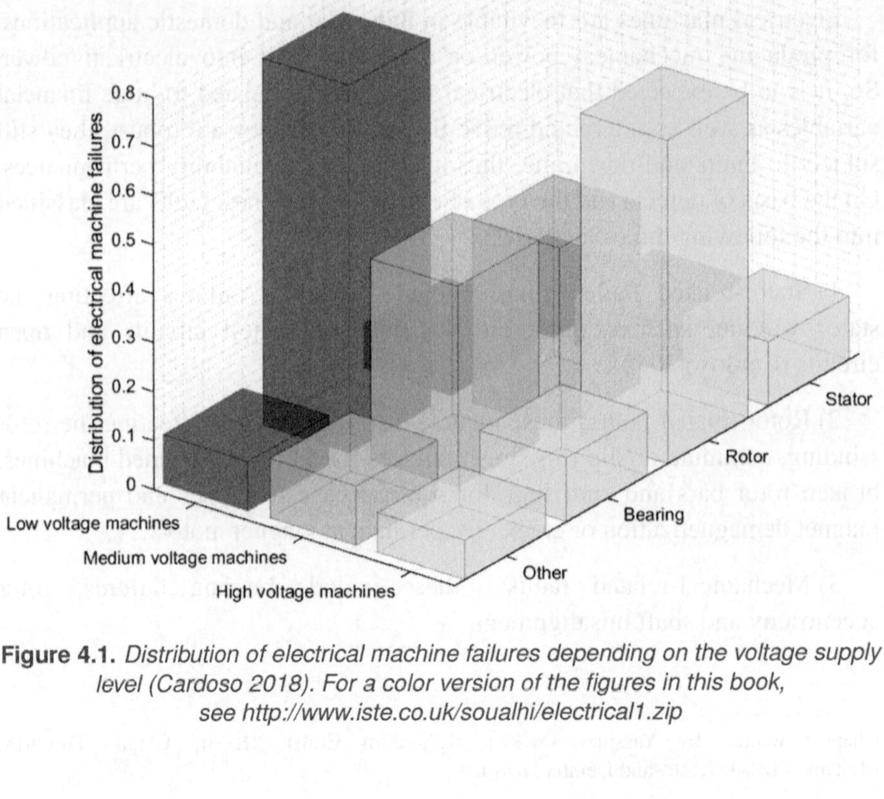

Figure 4.1. *Distribution of electrical machine failures depending on the voltage supply level (Cardoso 2018). For a color version of the figures in this book, see http://www.iste.co.uk/soualhi/electrical1.zip*

4.2. Condition monitoring of electrical machines

Condition monitoring of electrical machines is a broad scientific area, the ultimate purpose of which is to ensure the safe, reliable and continuous operation of electrical machines. Fault detection in electrical machines is certainly the most important key in the reduction of maintenance cost. Despite the availability of methods with vast experience in electrical machines, fault detection is still a challenging task, because induction machines are widely used in variable-speed drives and in renewable energy conversion systems. Thorsen and Dalva (1999) reviewed many tools and techniques developed for condition monitoring of electrical machines in order to prolong their lifespan, including existing and pre-installed sensors, such as those sensing speed, torque sensors, vibrations, temperature, flux density, and so on. These sensors are combined together in different architectures and coupled with algorithms to enable efficient monitoring of the condition of the system. Since the number of different machines, drives and applications are high, it is quite obvious that the number of electrical machine faults and diagnostic methods are also high. Among them, the motor current signature analysis (MCSA), analysis of the stator current harmonics index, is the most favorable (Thomson and Stewart 1988, Kliman *et al.* 1988). Most define the MSCA as the monitoring and spectral analysis of the stator current at a steady state. Despite its origin, its name is very generic, and it also analyzes the stator current spectra under transient operation. Anyway, this method has become favorable due to its unique characteristics such as remote monitoring (Antonino-Daviu *et al.* 2015), low implementation and equipment costs as well as continuous and online monitoring capability. The advantage of signature analysis of the motor electrical quantities is that it is a noninvasive technique, as these quantities are easily accessible during operation (Frosini *et al.* 2015). Moreover, stator currents are generally available for purposes such as control and protection as well, by avoiding the use of extra sensors (Seera and Lim 2014). Hence, most of the recent research detecting induction machine faults have been focused on electrical monitoring with an emphasis on current analysis (Leite *et al.* 2015, Benbouzid 2000).

4.2.1. *Detection of faults using MCSA*

Results from previous studies and a survey of condition monitoring for induction motors show important features of failure rates. The major faults of electrical machines can be broadly classified into the following (Thorsen and Dalva 1999, Zhang *et al.* 2011):

– static and/or dynamic air-gap irregularities;

– broken rotor bar or cracked rotor end rings;

– stator faults (opening or shorting of one or more coils of a stator phase winding);

– abnormal connection of the stator windings;

– bent shaft (akin to dynamic eccentricity) leads to friction between the rotor and the stator, causing serious damage to the stator core and windings;

– bearing and gearbox failures.

The most common faults are bearing faults, stator faults, rotor faults and eccentricity or any combination of these faults. When analyzed statistically, about 40% of the faults correspond to bearing faults, 30–40% to stator faults, 10% to rotors faults and the remaining 10% belong to a variety of other faults. The frequencies induced by each fault depend on the particular characteristic data of the motor (like synchronous speed, slip frequency and pole-pass frequency) as well as on the operating conditions.

4.2.2. *Bearing failures*

Since induction machine rotors are subjected to high stresses, for example, thermal stresses, mechanical stresses and electrical stresses, they are always more vulnerable compared to the stator. Particularly, bearings are more vulnerable to failure. Moreover, in many industries, bearing failures have been a persistent problem which accounts for a significant proportion of all failures in electrical machines; for example, the bearing failure in wind energy conversion systems generators is the most common failure associated with a long downtime. Bearing failure is typically caused by some misalignment in the drive train, which gives rise to abnormal loading and accelerates bearing wear, improper lubrication and occasionally manufacturing faults in the bearing components. A schematic view of a typical rolling-element bearing is shown in Figure 4.2.

Because of their construction, rolling-element bearings generate precisely identifiable signature on vibration (Wang and Jianu 2010, Tavner *et al.* 2008).

Those frequencies present an effective route for monitoring progressive bearing degradation.

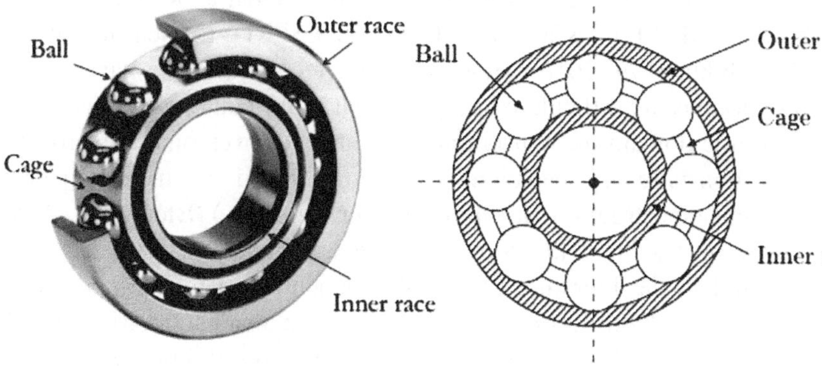

Figure 4.2. *Typical rolling-element bearing*

It is therefore possible to detect on the stator side the frequencies associated with the bearings using an accelerometer mounted directly on the bearing housing, which is not often easily accessible (Tavner *et al.* 2008). It is also true that vibration monitoring has make out its efficiency and is highly suitable for rolling-element bearings. However, it has problems when dealing with a good vibration baseline (Schoen *et al.* 1995). If no baseline is available, no history has been built-up, the background noise rises, and then it will be impossible to detect specific faults (Tavner *et al.* 2008). To tackle this problem, many works have proposed an alternative procedure for bearing wear detection in electrical machines by analyzing the stator side electrical quantities, such as the current (Schoen *et al.* 1995, Frosini *et al.* 2010), or the instantaneous power factor (Ibrahim *et al.* 2008). Indeed, a bearing fault is assumed to produce an air-gap eccentricity (Blodt *et al.* 2009) and consequently an unbalanced magnetic pull. Hence, this gives rise to torque oscillations, which lead to amplitude and/or phase modulation of the stator current (Blodt *et al.* 2009, Schoen *et al.* 1995, Riley *et al.* 1998). Therefore, it is sufficient to demodulate the current for bearing faults detection. So, the question is, "Which demodulation technique we have to investigate?", and a brief survey allowed to establish a roadmap for different demodulation techniques (Elbouchikhi *et al.* 2015*a*). For steady-state operations, current spectral estimation based on the fast Fourier transform (FFT) and its extension, the short-time Fourier transform (STFT), has been widely employed, such as FFT-based bispectrum/bicoherence (Benbouzid 2000). Since these techniques have frequency resolution limitation (Elbouchikhi *et al.* 2015*b*), high-resolution techniques are introduced, such as MUltiple SIgnal Classification (MUSIC) (Kia *et al.* 2007) and ESPRIT (Stoica and Moses 1997-Elbouchikhi *et al.* 2016). However, these techniques also have several drawbacks, as they are difficult to interpret and it is difficult to extract variation features in time domain for nonstationary signals. To overcome this problem and under nonstationary behavior, procedures based on time–frequency representations (spectrogram, quadratic Wigner Ville, etc.) (Blodt *et al.* 2008, 2009, Antonino-Daviu *et al.* 2009) or time-scale analysis (wavelet) have been proposed in the literature of electrical machines (Cusido *et al.* 2008, Riera-Guasp *et al.* 2008, Kia *et al.* 2009). There are also parametric methods based on parameter estimation of a known model (Elbouchikhi *et al.* 2015*b*). Nevertheless, these methods are formulated through integral transforms and analytic signal representations (Mandic *et al.* 2013), so their accuracy depends on data length, stationarity and model accuracy.

4.3. Signal processing tools

Most of electrical machine faults lead to current modulation (amplitude and/or phase) (Tavner 2008). This is the particular case of bearing faults (Stack *et al.* 2004). So, for failure detection, a possible approach relies on the use of amplitude demodulation techniques. The most investigated and popular amplitude demodulation techniques include Hilbert transform (HT), Teager energy operator TKEO (Kamwa *et al.* 2011) and synchronous demodulation. Furthermore, for a three-phase system, three-phase transformations such as Concordia transform (CT) (Trajin *et al.* 2009, Onel and Benbouzid 2008) and Park vector approach (Cruz and Cardoso 2001, Cardoso *et al.* 1999, Cardoso and Saraiva 1993) have been employed to perform demodulation. However, TKEO is less robust against noise, HT is valid only for mono-component signals, while CT and Park vector approaches are reliable only for balanced three-phase systems and hence require the use of three-phase currents. To go beyond these constraints, innovative techniques are investigated to track the fault component by source separation such as principal components analysis (PCA) (Choqueuse *et al.* 2012), noise cancellation (Zhou *et al.* 2008), and spectral features extraction and linear discriminant analysis (LDA) (Harmouche *et al.* 2015). Besides, in typical electrical machines,

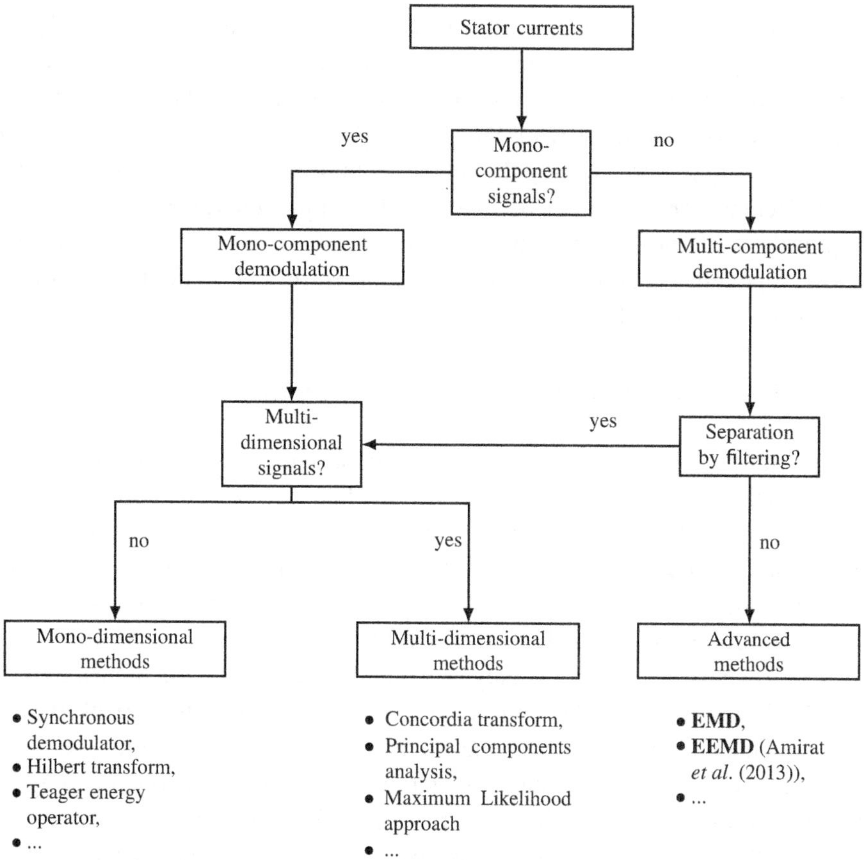

Figure 4.3. *Choosing a demodulation technique*

stator current components are the supply fundamental, harmonics, additional components due to slot harmonics, saturation harmonics, other components from unknown sources such as environmental noise and design imperfection, and eventually effect introduced by bearing faults. In typical electrical machines, the stator current has multicomponents and can be expressed by a temporal model as

$$x(t) = \sum_{k=1}^{M} a_k(t)\sin(\phi_k(t)), \qquad [4.1]$$

with $a_k(t) = a_k(1 + m_{ka}\sin(2\pi f_{ka}t + \varphi_{ka})))$, and $\phi_k(t) = 2\pi f_k t + m_{kp}\sin(2\pi f_{kp}t + \varphi_{kp})$. Here m_{ka} and m_{kp} are the AM index and the PM index, respectively, that can be introduced by a fault as an AM/PM effect. This work considers only the AM effect. Therefore, $m_{kp} = 0$ and $\phi_k(t) = 2\pi f_k t$, where $f_k = kf_0$, with f_0 being the fundamental frequency and k the harmonic order. Hence, for fault detection, a possible approach relies on the use of amplitude demodulation techniques to extract fault-related features. In this multicomponent signal context, the ensemble empirical mode decomposition (EEMD) is considered.

4.3.1. *Empirical mode decomposition method*

The EMD is an emerging signal processing algorithm for signal demodulation. It has been first introduced in Huang *et al.* (1998) and has since become an established tool for the analysis of nonstationary and nonlinear data (Tanaka and Mandic 2007). This approach has focused considerable attention and has been widely used for rotating machinery fault diagnosis (Yu *et al.* 2005, Antonino-Daviu *et al.* 2009, Amirat *et al.* 2013). It is an adaptive time–frequency data analysis method for nonlinear and nonstationary signals (Huang *et al.* 1998) and behaves like an adaptive filter bank (Gilles 2013). Conversely to FFT or wavelets that decompose a signal into a series of *sine* functions or a scaled mother wavelet; the aim of the EMD is to decompose the multicomponent signal into a series of intrinsic mode functions (IMFs) based on the local characteristic time-scale of the signal. This decomposition is described as follows:

– Identification of all extrema of the logged current.

– Interpolation between minima (resp. maxima) ending up with some envelope $e_{\min}(n)$ (resp. $e_{\max}(n)$).

– Computation of the mean:

$$R(n) = \frac{e_{\min}(n) + e_{\max}(n)}{2}. \qquad [4.2]$$

– Extraction of the detail:

$$d_m(n) = i(n) - R(n). \qquad [4.3]$$

– Iteration on the residue $R(n)$.

In practice, this algorithm has been refined by a sifting process until the detail d_m can be considered as IMF (Huang *et al.* 1998). To illustrate the EMD concept, let us assume the synthesized signal $x_{\text{syn}}(t)$ given by

$$x_{\text{syn}}(t) = a_1 \sin(\omega_1 t) + a_2 \sin(\omega_2 t), \quad [4.4]$$

where a_1 and a_2 are the amplitudes of the first and second components, respectively, whereas ω_1 and ω_2 are pulsations of those components. Figure 4.4 presents the results of decomposition of $x_{\text{syn}}(t)$ through the EMD algorithm.

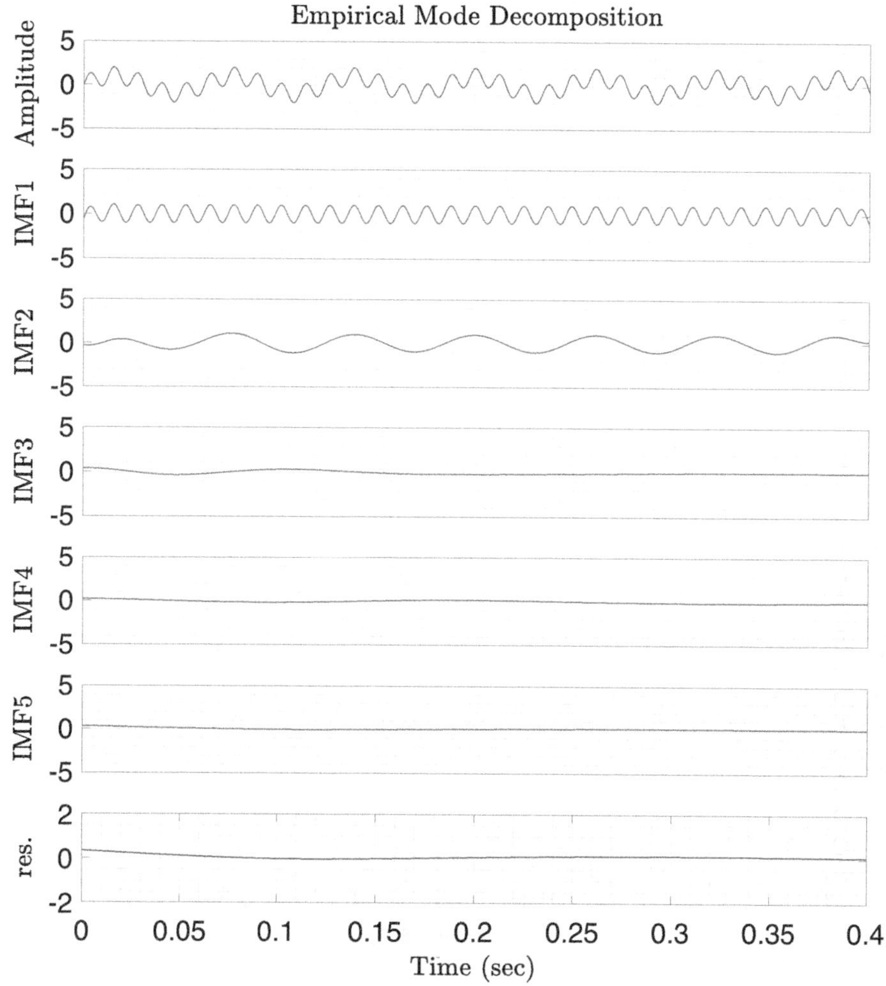

Figure 4.4. *EMD for uncorrupted synthetic signal*

It appears clearly that the two components are presented by the first and second IMFs. Unfortunately, real signals are not immunized from noises; and in order to have a look on the behavior of the EMD on the added noise signal, let us consider that the signal is corrupted by a white Gaussian noise (AWGN), then $x_{\text{syn}}(t)$ can be expressed by

$$x_{\text{syn}}(t) = a_1 \sin(\omega_1 t) + a_2 \sin(\omega_2 t) + \text{AWGN}, \qquad [4.5]$$

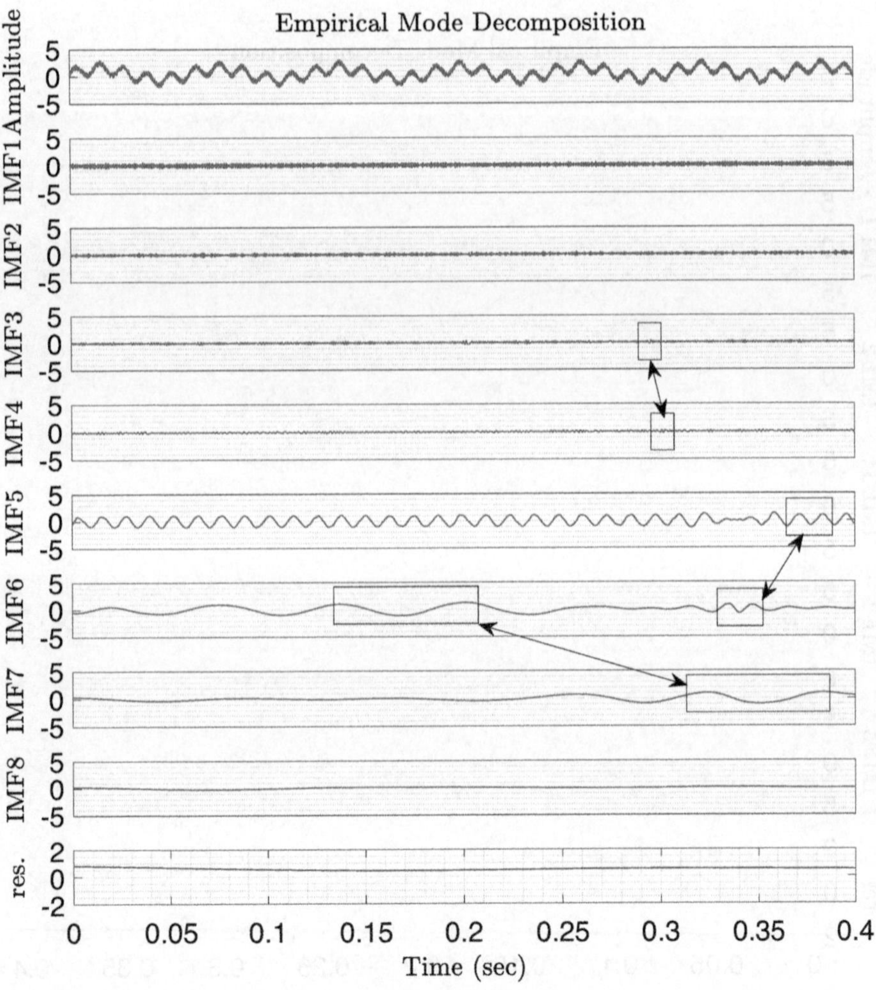

Figure 4.5. *EMD for a corrupted synthetic signal by an AWGN*

Figure 4.5 depicts the corresponding local time oscillations or IMFS and residue. The first remark is that the corresponding IMFs are shifted down from the fourth to seventh IMFs; this is due to the added noise to the original signal, hence high-frequency oscillations are introduced at the first, second and third IMFs. The second remark is the occurrence of the second component into at least two consecutive IMFs. This phenomenon is the mode mixing, as mentioned before. However, the main drawbacks EMD are: the ad-hoc process on which it is based (Gilles 2013), its sensitivity to noise and it suffers from mode mixing. Consequently, it is difficult to really understand what the EMD provides as IMFs are devoid of a physical meaning (Huang *et al.* 1998). To overcome the mode-mixing problem, the EEMD was introduced.

4.3.2. *EEMD principle*

As mentioned above, the main drawbacks of the EMD are: it is based on an ad-hoc process (Gilles 2013), it is mathematically difficult to model, it is noise sensitive, and it suffers from mode mixing. Consequently, it is difficult to understand what the EMD provides as IMFs that are devoid from a physical meanings (Huang *et al.* 1998). To deal with this drawback, an EEMD was proposed in Wu and Huang (2009) and Torres *et al.* (2011) and has become a tool for the analysis of nonstationary and nonlinear data (Tanaka and Mandic 2007) in a wide range of applications in signal processing (Huang and Shen 2014) and fault detection (Yu *et al.* 2005, Antonino-Daviu *et al.* 2009, Amirat *et al.* 2013). It is an improved EMD and is described as a noise-assisted data analysis method. Indeed, it deals with several EMD decompositions of the original signal corrupted by different artificial noises. The final EEMD is then the average of each EMD and defines true IMFs as the mean of an ensemble of trials. The EEMD algorithm is depicted in Figure 4.6 and its implementation is described step by step in Amirat *et al.* (2018).

The reliability of EEMD depends on the ensemble number M and the noise amplitude a. These two parameters are linked by (Wu and Huang 2009)

$$e = \frac{a}{\sqrt{M}}, \qquad [4.6]$$

where e is the standard deviation error, which is defined as the discrepancy between the input signal and the corresponding IMF.

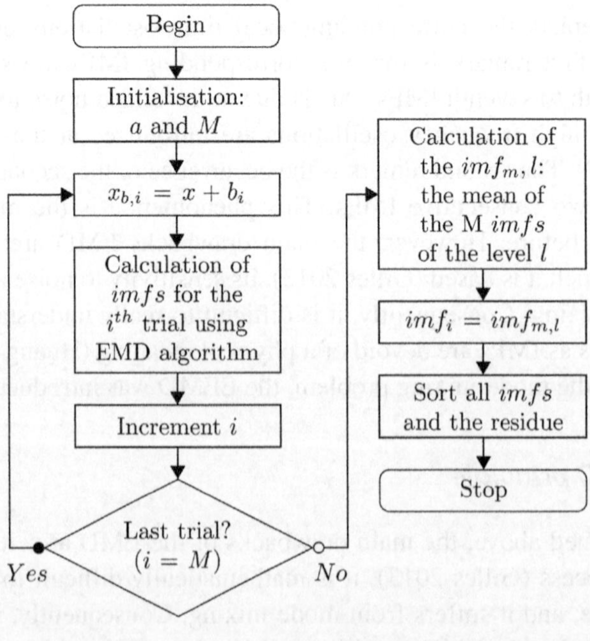

Figure 4.6. *EEMD process for signal decomposition*

So, through EEMD algorithm, a signal $x(t)$ can be expressed as a sum of k modes or IMFs as follow:

$$x(t) = \sum_{i=1}^{k} \text{IMF}_i(t) + \text{res}(t). \qquad [4.7]$$

Figures 4.7 and 4.8 illustrate the decomposition of free-noise signal and corrupted signal, respectively.

Let us denote by $x(n)$, $n = 1, \ldots, N$, the stator current. Under the multi-component assumption, the sampled current $x(n)$ can be decomposed as

$$x(n) = \sum_{i=1}^{j} \text{IMF}_i(n) + \text{res}(n), \qquad [4.8]$$

where $\text{IMF}_i(n)$ is the ith intrinsic mode function, res(n) is the residue and j the total number of IMFs. In practice, IMFs are unknown and must be extracted from the stator current $x(n)$. However, according to the EEMD principle, at

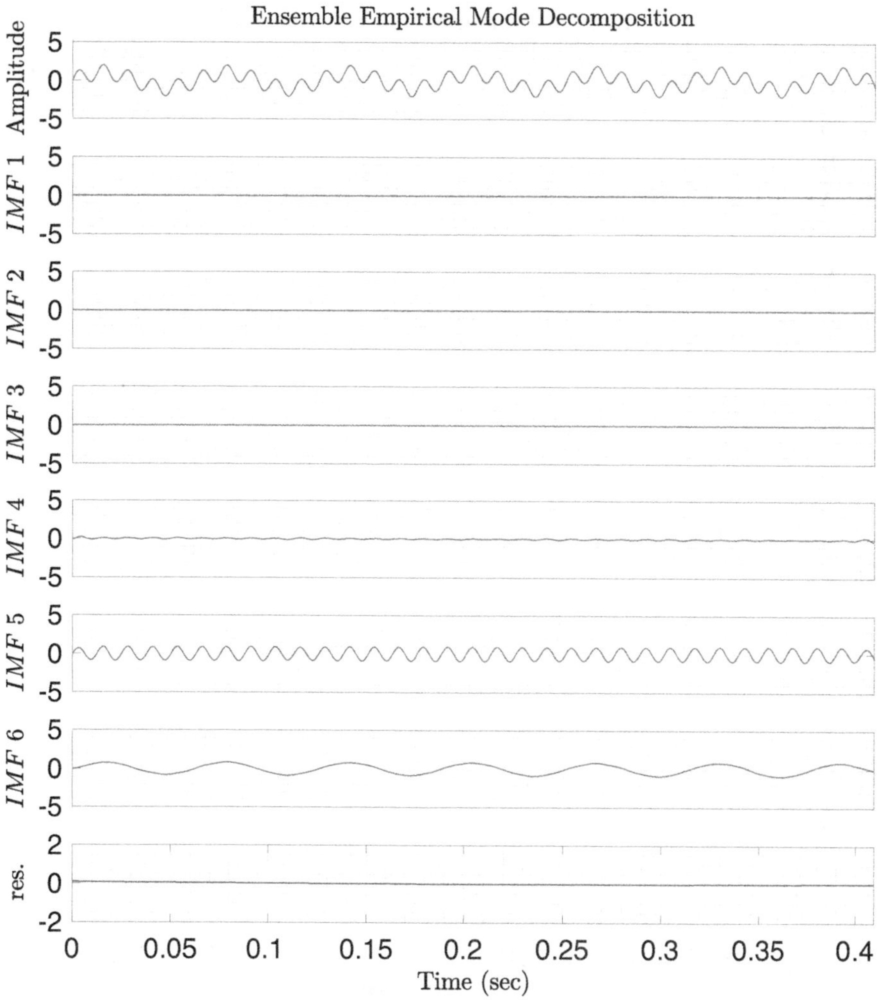

Figure 4.7. *EEMD for uncorrupted synthetic signal*

least one IMF is related or representative of the main component. hence, $x(n)$ can be expressed by

$$x(n) = \sum_{i=1}^{c-1} \mathrm{IMF}_i(n) + \mathrm{IMF}_c(n) + \sum_{i=c+1}^{j} \mathrm{IMF}_i(n) + \mathrm{res}(n), \qquad [4.9]$$

where $\mathrm{IMF}_c(n)$ is the closest intrinsic mode function to the original signal $x(n)$.

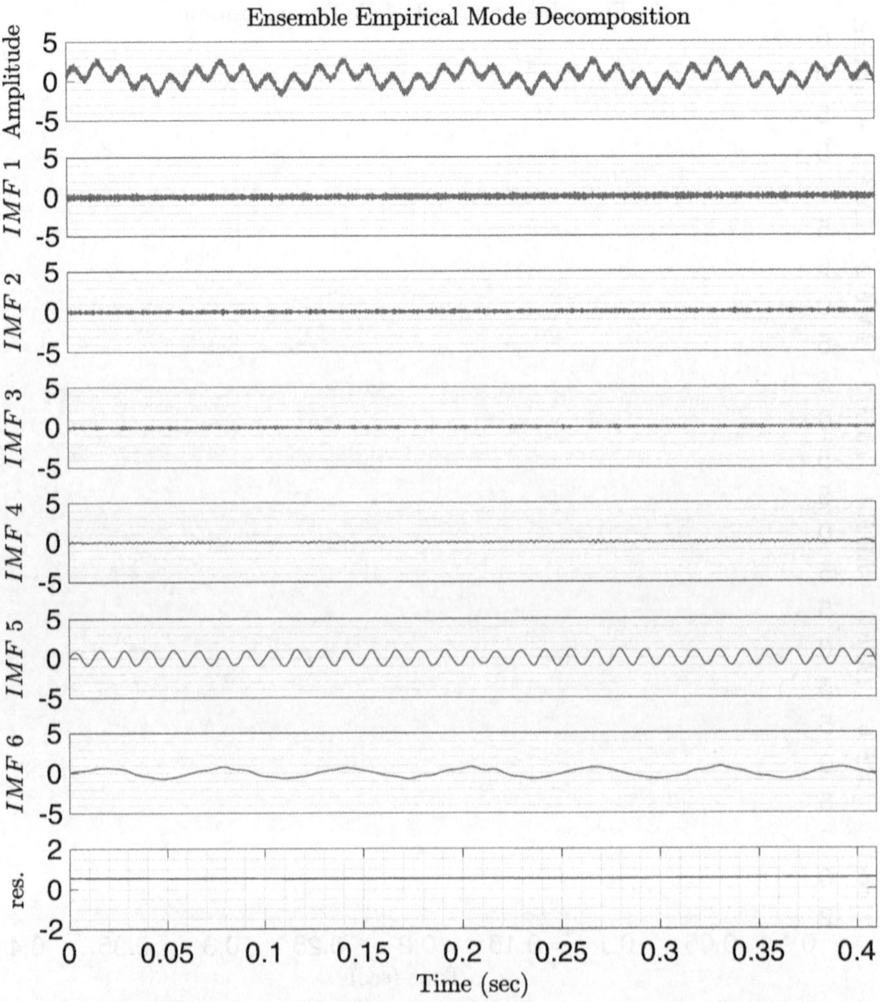

Figure 4.8. *EEMD for a corrupted synthetic signal by an AWGN*

4.3.3. *Variational mode decomposition principle*

The variational mode decomposition is an emerging signal processing algorithm and has been proposed by Dragomiretskiy and Zosso (2014) to answer the lack of theoretical background of the EMD algorithm. It analyzes nonstationary and nonlinear data and was investigated in various fields such as in crude oil risk forecasting (He *et al.* 2018) and bearing fault detection

(Chen *et al.* 2019, Wang *et al.* 2019, Jiang *et al.* 2018). The VMD is considered as a self-adaptive and quasi-orthogonal and entirely nonrecursive signal processing method, unlike the EMD/EEMD which are self-adaptive but recursive methods. The core of the VMD algorithm is based on classical Wiener filtering, Hilbert transform and heterodyne demodulation.

Like EEMD, a signal $x(t)$ can be decomposed by the VMD algorithm into k modes or subsignals $u_k(n)$, and each mode is compact around a center pulsation $\omega_k(t)$ (Chen *et al.* 2019), and $x(n)$ can also be expressed by [4.10] as follows:

$$x(t) = \sum_{i=1}^{k} \mathrm{Mode}_i(t) + \mathrm{res}(t). \qquad [4.10]$$

In addition, for the VMD algorithm, a constrained variational problem has to minimize the sum of each *Mode* component's bandwidth. The evaluation of the bandwidth can be implemented in the following three steps (Zhao and Li 2017):

– Step 1: Hilbert transform is applied to calculate the analytic signal of each Mode component, so that unilateral frequency spectrum can be obtained.

– Step 2: An exponential is used to modulate the estimated central frequency of each IMF component, thus frequency spectrum is shifted to the "baseband."

– Step 3: The H^1 Gaussian smoothness of the demodulated signal is calculated in order to estimate each IMF component's bandwidth.

Consequently, the mathematical model of constrained variational problem can be given as (Dragomiretskiy and Zosso 2014, Zhao and Li 2017):

$$\min_{\{u_k\},\{\omega_k\}} \left\{ \sum_{k=1}^{K} \left\| \partial_t \left\{ \left[\delta(t) + \frac{\mathrm{j}}{\pi t} \right] * u_k(t) \right\} \mathrm{e}^{-\mathrm{j}\omega_k t} \right\|_2^2 \right\}, \qquad [4.11]$$

s.t. $\sum_k \{u_k(t)\} = x(t)$, where $u_k = \{u_1, u_2, \ldots, u_k\}$ represents the modal function vector, $\omega_k = \{\omega_1, \omega_2, \ldots, \omega_k\}$ is the central frequency vector, ∂_t is the partial derivative of the function for time t, $\delta(t)$ is the Dirac function, j is the imaginary unit and $*$ is the convolution operator. To resolve the constrained problem, it must be transformed into an unconstrained problem, and for this purpose a quadratic penalty factor η and Lagrangian multiplier $\lambda(t)$ are introduced in Dragomiretskiy and Zosso (2014), and hence a new unconstrained variational problem expression is obtained as follows:

$$\begin{aligned}\Gamma(u_k, \omega_k, \lambda) = \eta \sum_{k=1} & \left\| \partial_t \left\{ \left[\delta(t) + \frac{\mathrm{j}}{\pi t} \right] * u_k(t) \right\} \mathrm{e}^{-\mathrm{j}\omega_k t} \right\|_2^2 \\ & + \left\| x(t) - \sum u_k(t) \right\|_2^2 \\ & + \left\langle \lambda(t), x(t) - \sum u_k(t) \right\rangle . \end{aligned} \qquad [4.12]$$

To resolve this unconstrained problem, we investigate the algorithm developed in Dragomiretskiy and Zosso (2014), where its straightforward implementation is described step by step in Algorithm 1.

Algorithm 1 VMD

Initialize:
    $\{u_k^1\}, \{\omega_k^1\}, \lambda^1, n \leftarrow 0$
repeat
    $n \leftarrow n + 1$
    for $k = 1$ to K **do**
        Update u_k
        $u_k^{n+1} \leftarrow \underset{\{u_k\}}{\arg\min}\, \mathcal{L}\left(\{u_{i<k}^{n+1}\}, \{u_{i\geq k}^{n+1}\}, \{\omega_i^n\}, \{\lambda^n\}\right)$
    end for
    for $k = 1$ to K **do**
        Update ω_k
        $\omega_k^{n+1} \leftarrow \underset{\{\omega_k\}}{\arg\min}\, \mathcal{L}\left(\{u_i^n\}, \{\omega_{i<k}^{n+1}\}, \{\omega_{i\geq k}^{n+1}\}, \{\lambda^n\}\right)$
    end for
    Dual ascent:
    $\lambda^{n+1} \leftarrow \lambda^n + \tau\left(f + \sum_k u_k^{n+1}\right)$
until convergence: $\sum_k \|u_k^{n+1} - u_k^n\|_2^2 / \|u_k^n\|_2^2 < \epsilon$

Figures 4.9 and 4.10 illustrate the decomposition of free-noise signal and corrupted signal, respectively.

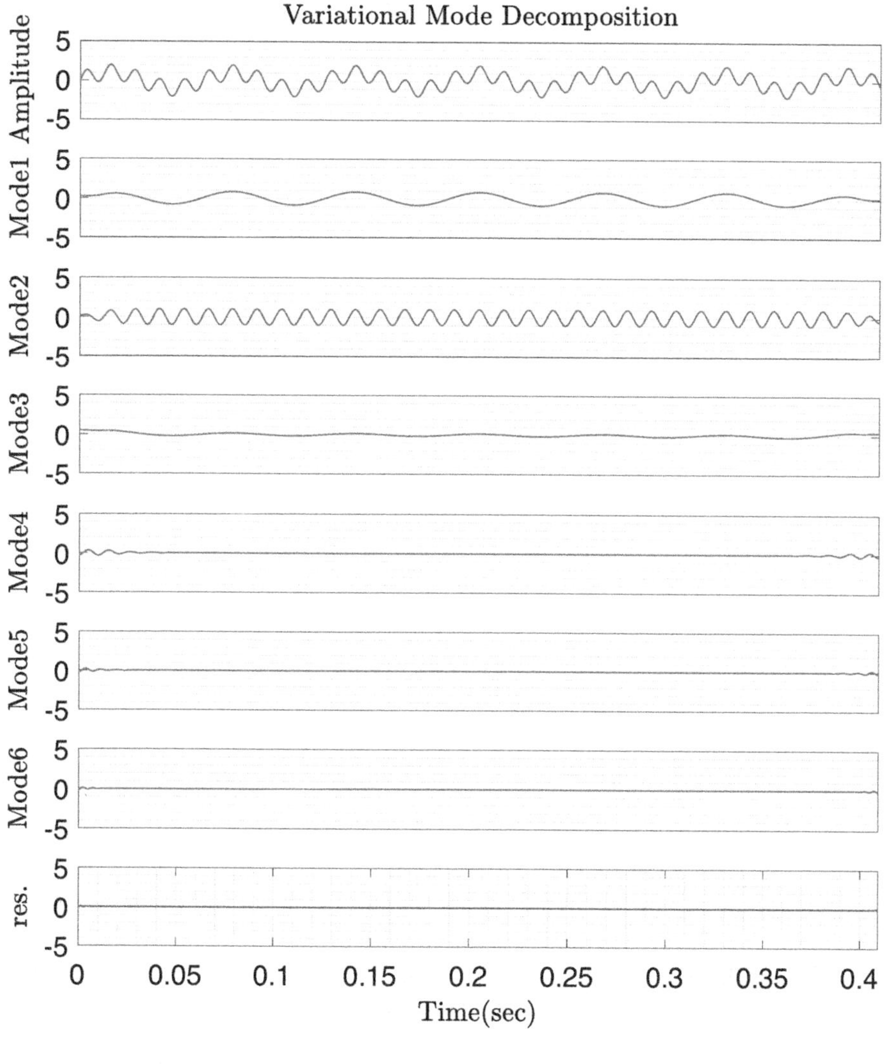

Figure 4.9. *VMD for uncorrupted synthetic signal*

It is noted that the components obtained through both EEMD and VMD algorithms highlight the presence of a closest IMF (or Mode), so, the main issue that rises is how to extract this IMF (or Mode)? In this context, a mode decomposition-based notch filter was developed in Amirat *et al.* (2016) and

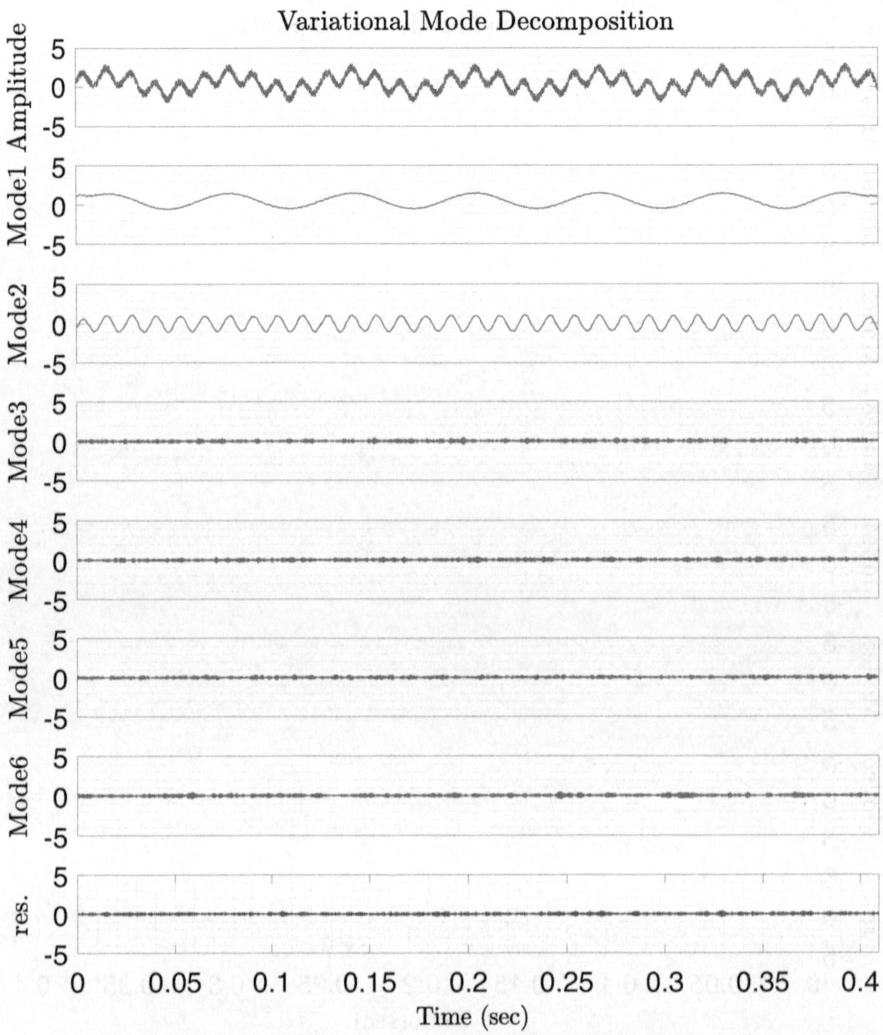

Figure 4.10. *VMD for a corrupted synthetic signal by an AWGN*

Amirat *et al.* (2019). The aim of the notch filter is to cancel the dominant IMF or Mode, and the result denoted $x(n)_{d_{\mathrm{EEMD}}}$ and $x(n)_{d_{\mathrm{VMD}}}$ can therefore be used to detect bearing failures.

4.4. Mode decomposition-based notch filters

As mentioned in previous sections, the decomposition of signal $x(t)$ through EEMD and VMD leads to a sum of modes or subsignals as expressed in [4.13] and [4.14], respectively. Among these modes, at least one mode is closely representative to the original signal, which is the dominant mode denoted $\mathrm{Mode_d}(n)$. Assuming that the occurrence of a fault introduces a new component in the original signal, hence a specific mode or subsignal denoted Mode_e is introduced in the mode decomposition of this original signal. Therefore, the sampled signal $x(n), n = 1, \cdots, N$, can be decomposed through both algorithms as follows:

$$x(n) = \mathrm{IMF_d}(n) + \sum_{i=1}^{k<j} \mathrm{IMF}_i(n) + \mathrm{res}(n), \qquad [4.13]$$

$$x(n) = \mathrm{Mode_d}(n) + \sum_{i=1}^{k<j} \mathrm{Mode}_i(n) + \mathrm{res}(n). \qquad [4.14]$$

4.4.1. *Statistical distance measurement*

In statistics and information theories, a statistical distance quantifies the distance between two statistical quantities, which can be two random variables, or two probability distributions or samples. In statistics literature, various approaches have been indexed and investigated in various fields, particularly for fault detection and diagnostic (Harmouche *et al.* 2015, Amirat *et al.* 2018). The statistical tool known as Pearson correlation is used to measure the distance and to give a weight to dependency between two temporal series $x(n)$ and $y(n)$ (Proakis and Manolakis 1996). This dependency is weighted by a coefficient denoted $r(x, y)$ and defined by 4.15, a value of this coefficient close to -1 or 1, indicates that $x(n)$ and $y(n)$ are highly correlated positively or negatively, respectively, whereas a value around 0 indicates that there is no dependency between $x(n)$ and $y(n)$:

$$r(x, y) = \frac{\sum_n \left[(x(n) - m_x) * (y(n) - m_y)\right]}{\sqrt{\sum_n (x(n) - m_x)^2} . \sqrt{\sum_n (y(n) - m_y)^2}}, \qquad [4.15]$$

where m_x and m_y are the means of x and y, respectively.

4.4.2. *Dominant Mode cancellation*

The cancellation of the dominant Mode (i.e., IMF) is illustrated in Figure 4.11. The cancellation of the dominant IMF or Mode algorithm can be implemented in the following three steps:

– Step 1: The EEMD (i.e., VMD) is applied to calculate the set of modes (i.e., IMFs) contained in the analyzed signal.

– Step 2: Pearson correlation coefficient is calculated using 4.15 as many times as there are modes or IMFs, then $r_{\mathrm{d}} \approx 1$ indexes the $\mathrm{IMF_d}$ (i.e., $\mathrm{Mode_d}$).

– Step 3: Then, the indexed $\mathrm{IMF_d}$ or $\mathrm{Mode_d}$ is subtracted from the analyzed signal $x(n)$ and the result denoted $x(n)_d$EEMD and $x(n)_d$VMD can therefore be used to detect bearing failures.

As mentioned above, Pearson correlation is a measurement of the strength of the association between two variables.

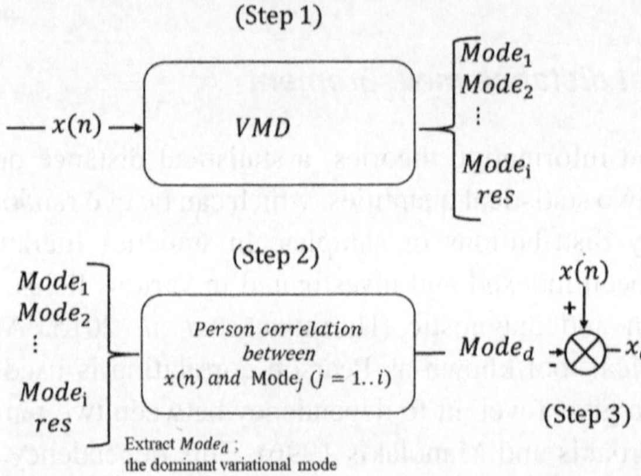

Figure 4.11. *Dominant Mode cancellation (Amirat* et al. *2018)*

Let us consider $X(n)$ as line current and $Y(n)$ as the IMF or Mode. So,

$$X(n) = x(n), \tag{4.16}$$

and

$$Y_i(n) = \mathrm{IMF}_i(n), \tag{4.17}$$

or

$$Y_i(n) = \text{Mode}_i(n), \quad [4.18]$$

where $(i = 1, \ldots, j)$ corresponds to the IMF or Mode rank. Hence, the Pearson correlation coefficient r_i is computed for each pair $(X(n), Y_i(n))$, and then $r_\text{d} \approx 1$ indexes the IMF_d. Once the IMF_d (or Mode) is determined, it is canceled from the original signal $x(n)$, and the remaining signal $x_\text{c}(n)$ expressed by (4.19) for EEMD and by (4.20) for VMD can therefore be investigated for bearing failure detection.

$$x_{\text{c}_\text{EEMD}}(n) = x(n) - \text{IMF}_\text{d}(n), \quad [4.19]$$

$$x_{\text{c}_\text{VMD}}(n) = x(n) - \text{Mode}_\text{d}(n). \quad [4.20]$$

The cancellation process is repeated until there is no correlation between the main signal $x(n)$ and the IMFs (or Modes) contained in $x_\text{c}(n)$. This signal can therefore be investigated for bearing faults detection.

4.5. Fault detectors

For fault detection, several detectors based on the instantaneous amplitude have been proposed in the literature. These include hidden Markov model (Ocak and Loparo 2001, Miao and Makis 2007), artificial neural networks (Nejjari and Benbouzid 2000), fuzzy logic (Onel and Benbouzid 2008) and support vector machines (Guo *et al.* 2009). However, most of these approaches use unnecessary and complicated classifiers and assume that a training database is available. This can be very difficult to obtain for many industrial applications. Indeed, it has been mentioned in a number of previously published papers that one of the main difficulties in real-world testing of developed condition monitoring technique is the lack of collaboration with industrial operators and manufacturers due to data confidentiality, particularly when failures are present (Yang *et al.* 2010) and is difficult to obtain (Kusiak 2016). In this section, we propose a low-complexity detector, based on the statistical features and that does not require any training. For this purpose, several statistical parameters were used to define signals and time-series average properties. The two basic parameters are the mean value μ and the standard deviation σ. For a given data set $x(i)$, with $i = 1, \ldots, N$, these are defined as follows:

$$\mu = \frac{1}{N} \sum_{n=1}^{N} x(n), \qquad [4.21]$$

$$\sigma = \sqrt{\frac{1}{N} \left(\sum_{n=1}^{N} (x(n) - \mu)^2 \right)}. \qquad [4.22]$$

It is therefore proposed to compute the variance of $x_c(n)$ to detect a fault. This statistical criterion, denoted σ^2, measures the deviation of the amplitude around its mean μ. It is given by

$$\sigma^2 = \frac{1}{N} \sum_{n=0}^{N-1} (x_c(n) - \mu)^2, \qquad [4.23]$$

where μ is the average of $x_c(n)$. To avoid the EEMD/VMD edge effect problem, $x_c(n)$ is then truncated by removing α samples at the beginning and at the end of $x_c(n)$. Hence, the proposed criterion σ^2 is expressed by (Amirat *et al.* 2010):

$$\sigma^2 = \frac{1}{N - 2\alpha} \left(\sum_{n=\alpha}^{N-\alpha-1} (x_c(n) - \mu)^2 \right), \qquad [4.24]$$

and

$$\mu = \frac{1}{N - 2\alpha} \sum_{n=\alpha}^{N-\alpha-1} x_c(n). \qquad [4.25]$$

The hypothesis to detect a fault can therefore be formulated as follows.

If $\sigma^2 > \gamma$, the machine is faulty, where γ is a threshold. For ideal acquisition conditions $\gamma = 0$, but in real-world applications, there is always an added noise to measurements, then γ can be set subjectively.

4.6. Simulation and experimental validation

4.6.1. *Synthetic signals*

In this first validation step, simulations have been performed with AM synthetic signals. According to [4.1], and since additional components could be considered as noise in the context of bearing faults detection (Zhou *et al.* 2008), the AM synthetic signal corrupted by an additive noise δ is defined as

$$x(n) = \underbrace{(1 + \beta \sin(\omega_2 n + \psi))}_{a(n)} \cos(\omega n + \phi) + \delta(n), \qquad [4.26]$$

where $n = 0, \ldots, N-1$ is the sample index, N is the number of samples and ϕ is the phase parameter. In [4.26], frequency ω is equal to $2\pi f/F_e$ and ω_2 is equal to $2\pi f_2/F_e$ (where f, f_2 and F_e are the supply, fault and sampling frequency, respectively) and amplitude $a(n)$ is related to the fault. It should be noted that the additive noise $\delta(n)$ is supposed to be a zero mean and Gaussian noise process. This assumption is an approximation of the electrical noise picked up in the wiring and signal conditioning circuits (Phadke 2002) and is widely considered in the measurement and electrical engineering communities (Komaty *et al.* 2014). The modulation index β is the fault index; $\beta = 0$ is for the healthy case and $\beta > 0$ is for the faulty one. Simulations have been carried out with a sampling frequency $F_e = 10\,\text{kHz}$, a supply frequency $f = 50\,\text{Hz}$ and $f_2 = 100\,\text{Hz}$. In order to simulate healthy and faulty cases, the modulation index has been set, respectively, to $\beta = 0.0$ for healthy cases and $\beta = 0.1, 0.15$ and 0.2 for different fault severities (Figure 4.12).

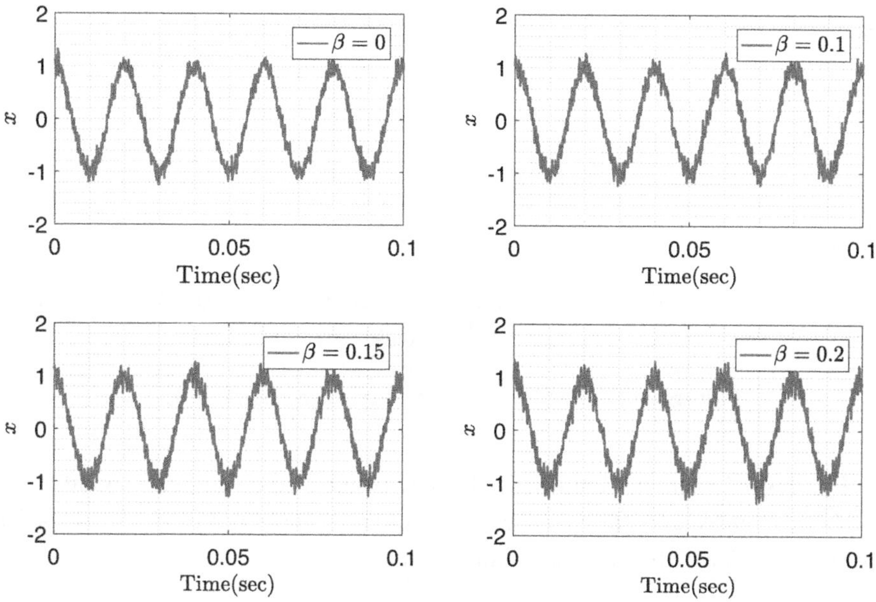

Figure 4.12. *Time representation of the synthetic signal for different modulation index*

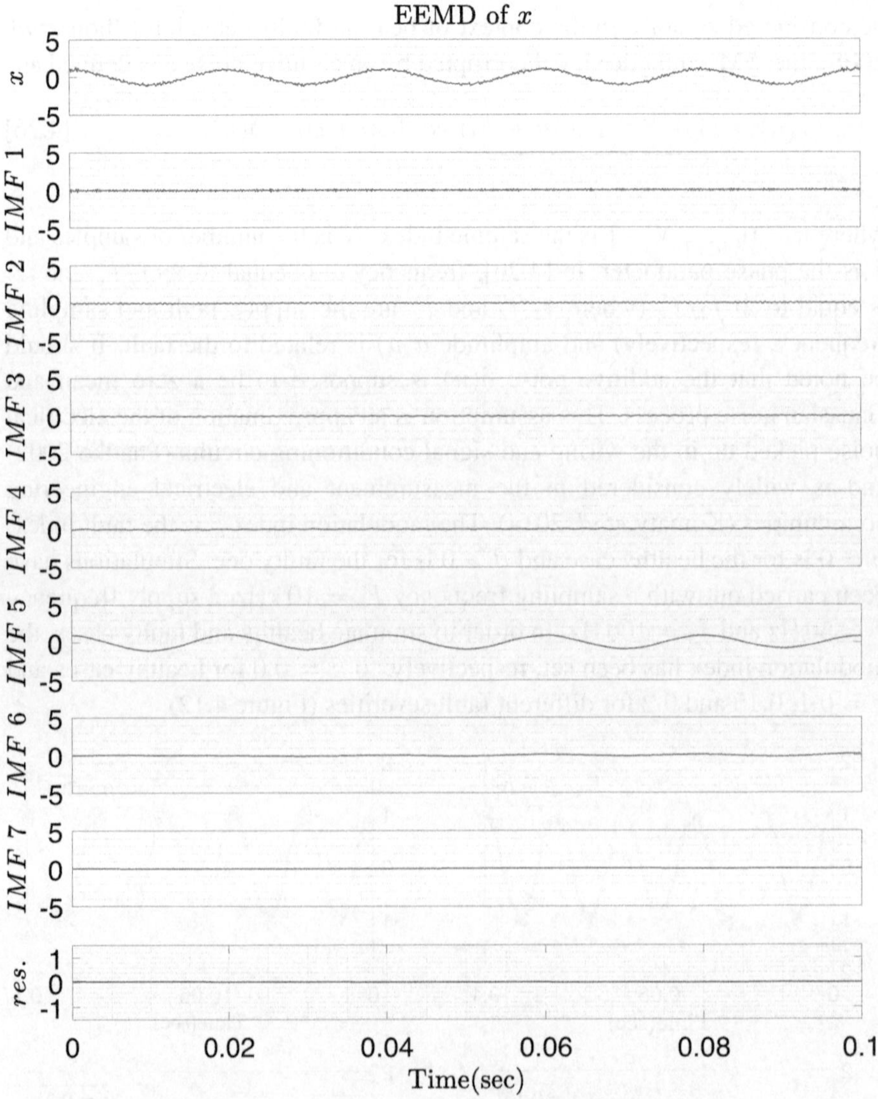

Figure 4.13. *EEMD for modulated synthetic signal:* $\beta = 0$

Figures 4.13 and 4.14 show the EEMD result of the synthetic signal $x(n)$ for both healthy and faulty cases, respectively, and Figures 4.15 and 4.16 illustrate the VMD result of the same signals.

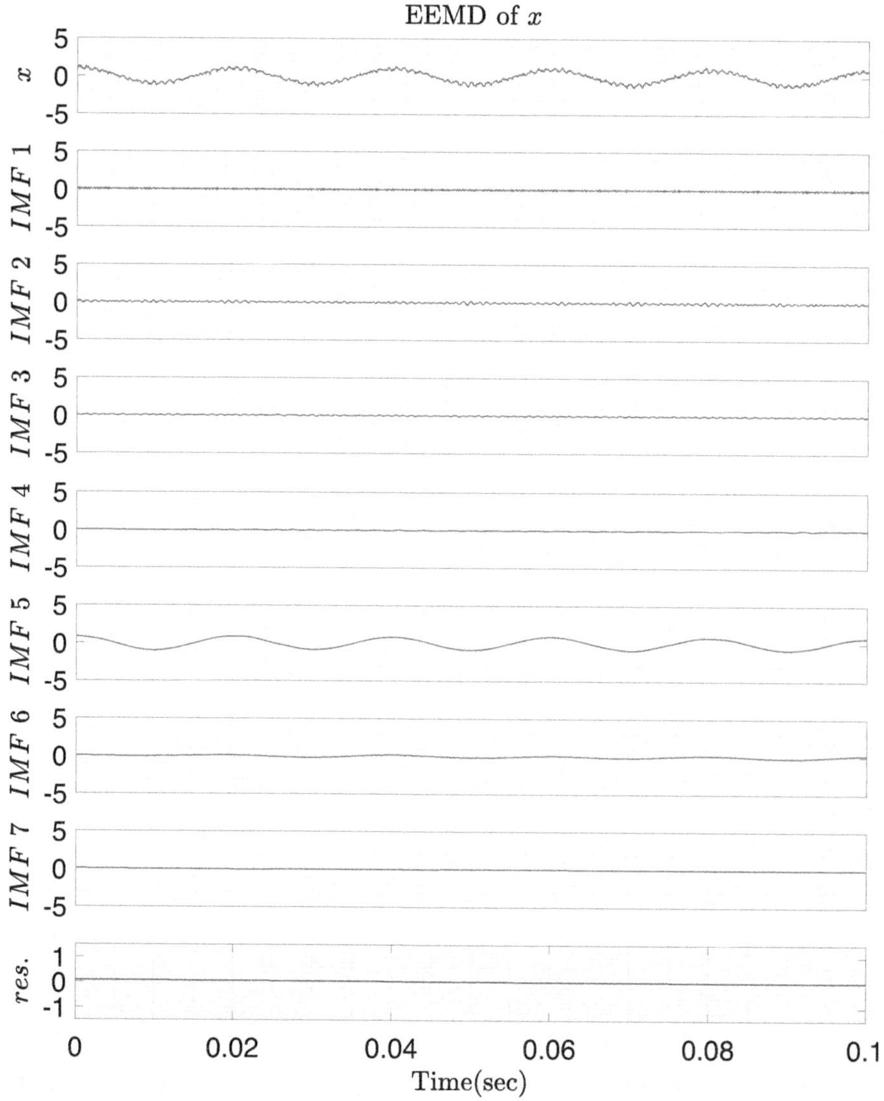

Figure 4.14. *EEMD for modulated synthetic signal:* $\beta = 0.2$

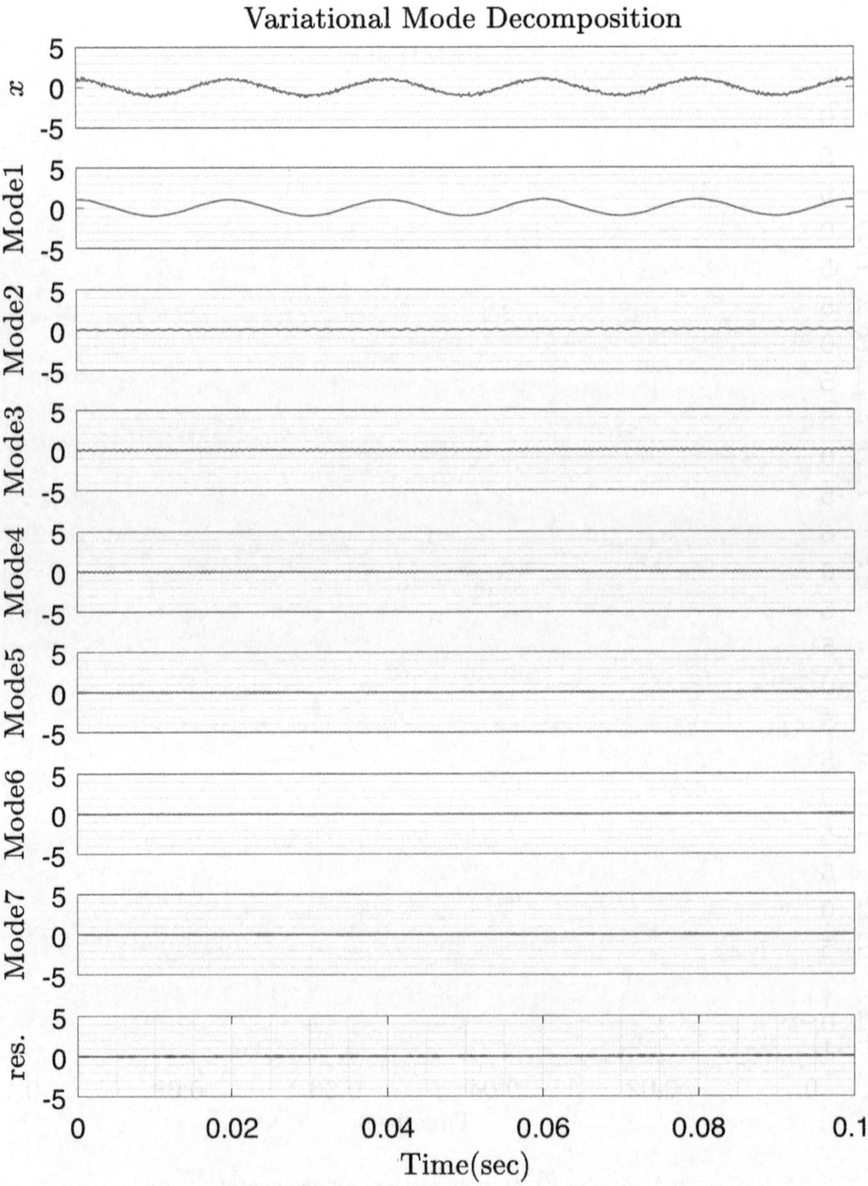

Figure 4.15. *VMD for modulated synthetic signal:* $\beta = 0$

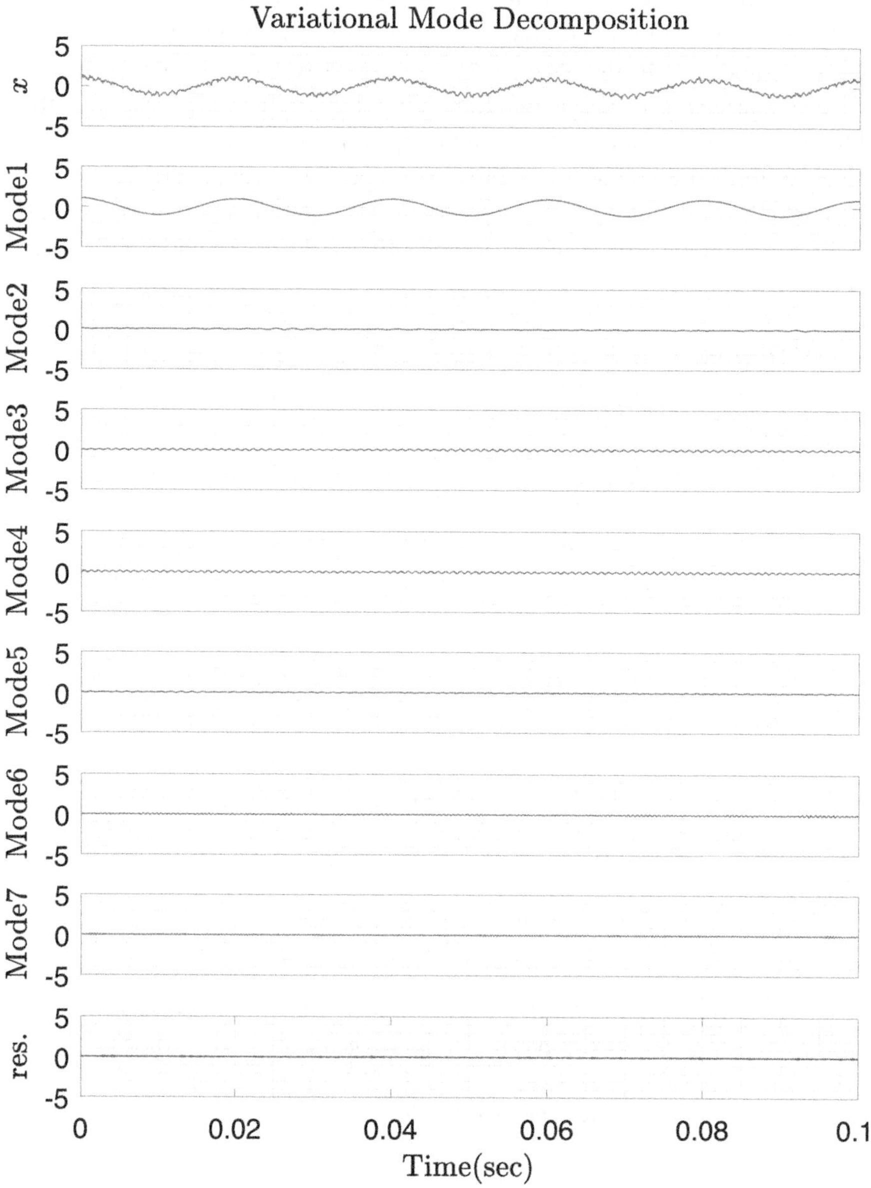

Figure 4.16. *VMD for modulated synthetic signal:* $\beta = 0.2$

It clearly shows that at least one IMF or Mode is close to the original signal. In order to quantify the strength of the association between $x(n)$ and each IMF, the Pearson correlation coefficient r_i is computed and results are depicted in Tables 4.1 and 4.2. In this case, IMF_5 is the closest to the main signal for the EEMD case and Mode_1 is the dominant mode in the VMD case. It is then subtracted, and the variance of the remaining signal is computed and results for both algorithms are presented in Tables 4.1 and 4.2, respectively. It is clearly shown that the fault criterion σ^2 rises with the modulation index β, as presented in Tables 4.3 and 4.4. For healthy case ($\beta = 0$) and due to the added noise $\delta(n)$, σ^2 is not equal to 0.

IMF rank	$\beta = 0.0$	$\beta = 0.1$	$\beta = 0.15$	$\beta = 0.2$
IMF_1	0.1129	0.1182	0.1189	0.1193
IMF_2	0.0873	0.1143	0.1324	0.1543
IMF_3	0.0688	0.0811	0.1141	0.1387
IMF_4	0.0598	0.0583	0.0284	0.0493
IMF_5	0.9883	0.9820	0.9825	0.9771

Table 4.1. *Pearson correlation coefficients of synthetic signal for EEMD*

IMF rank	$\beta = 0.0$	$\beta = 0.1$	$\beta = 0.15$	$\beta = 0.2$
Mode_1	0.9913	0.9883	0.9862	0.9816
Mode_2	0.0571	0.0541	0.0563	0.0530
Mode_3	0.0626	0.0890	0.0567	0.1090
Mode_4	0.0535	0.0556	0.1131	0.1123
Mode_5	0.0546	0.0609	0.0599	0.0572

Table 4.2. *Pearson correlation coefficients of synthetic signal for VMD*

$\beta = 0.0$	$\beta = 0.1$	$\beta = 0.15$	$\beta = 0.2$
$\sigma^2 = 0.0044$	$\sigma^2 = 0.0057$	$\sigma^2 = 0.0067$	$\sigma^2 = 0.0076$

Table 4.3. *The variance (σ^2) of x_c for EEMD*

$\beta = 0.0$	$\beta = 0.1$	$\beta = 0.15$	$\beta = 0.2$
$\sigma^2 = 0.0073$	$\sigma^2 = 0.0106$	$\sigma^2 = 0.0128$	$\sigma^2 = 0.0180$

Table 4.4. *The variance (σ^2) of x_c for VMD*

4.6.2. *Experimental validation*

In this second step, the proposed fault detector is experimentally tested on a test rig, and Figure 4.17 illustrates the basic layout of this test rig.

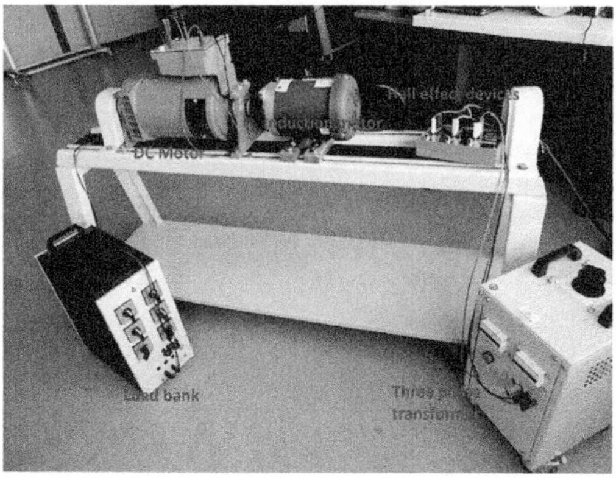

Figure 4.17. *Experimental setup*

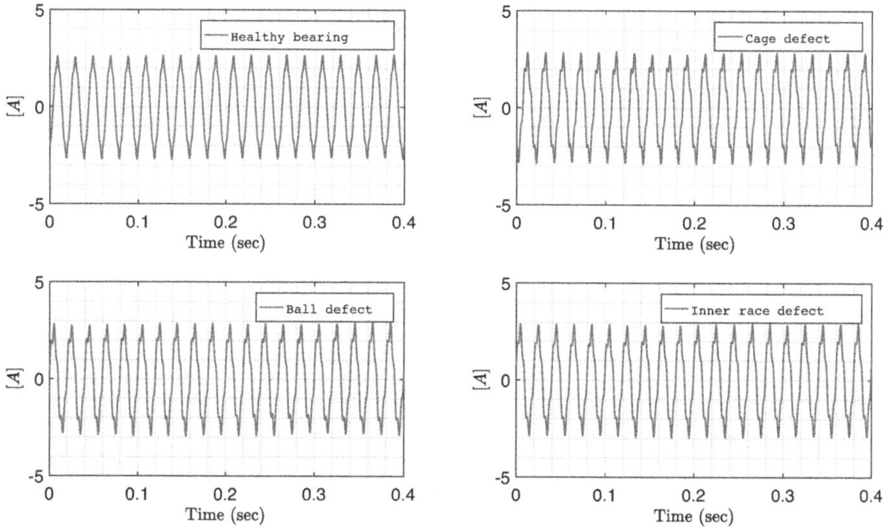

Figure 4.18. *Stator current $x(n)$*

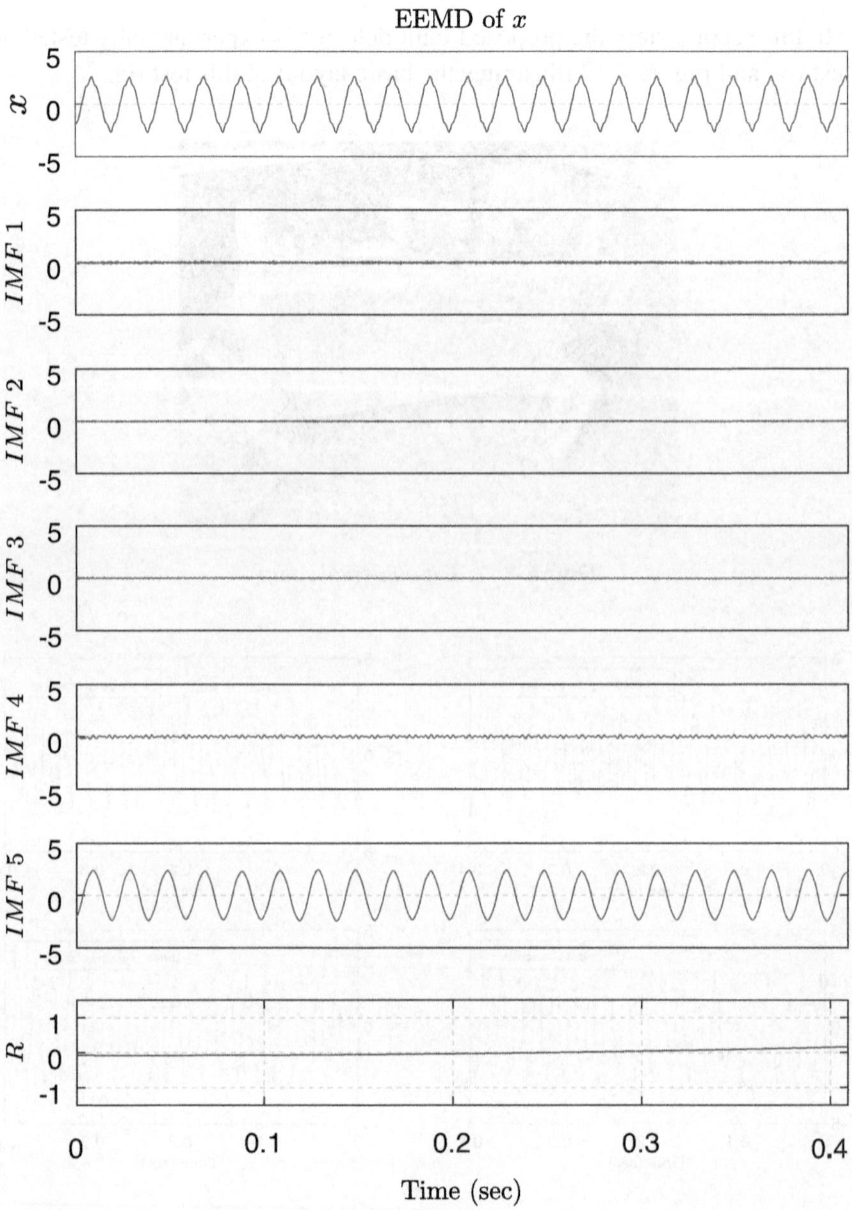

Figure 4.19. *EEMD decomposition of the stator current: Healthy bearing case*

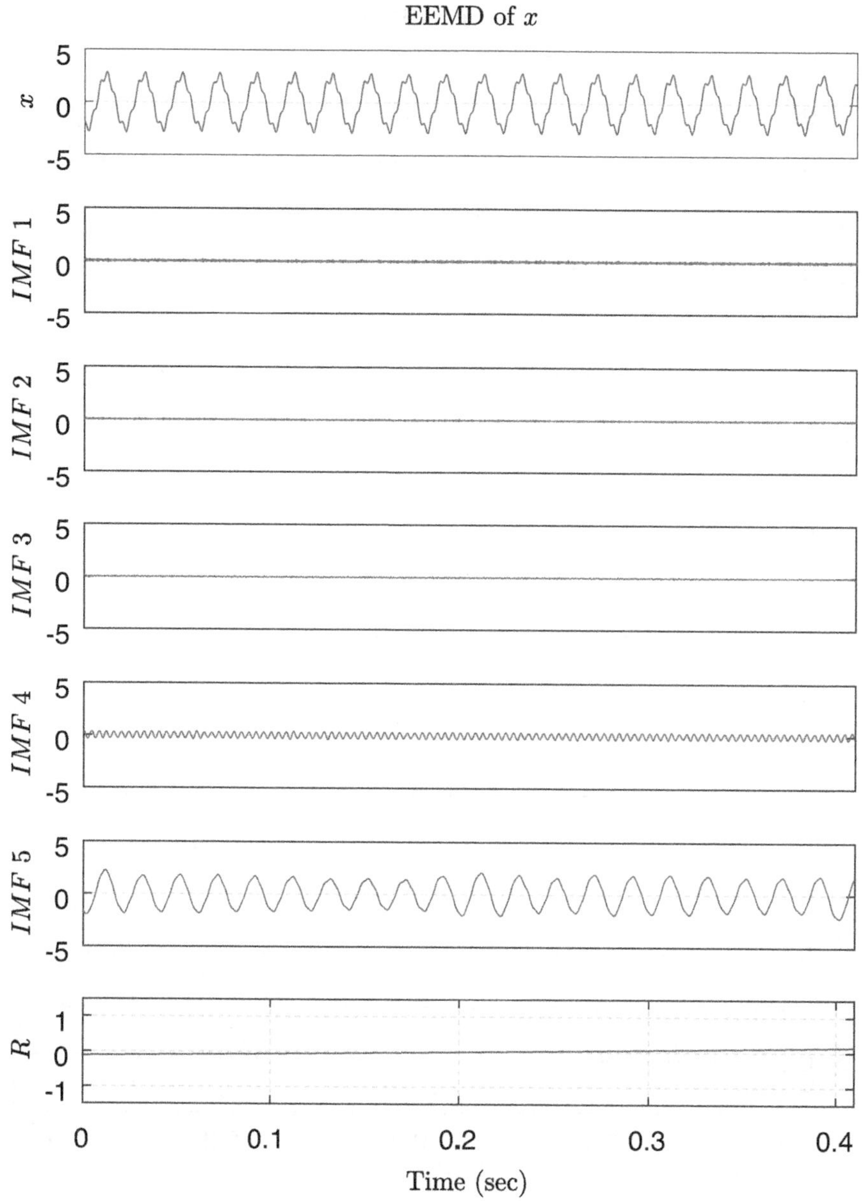

Figure 4.20. *EEMD decomposition of the stator current: Cage defect case*

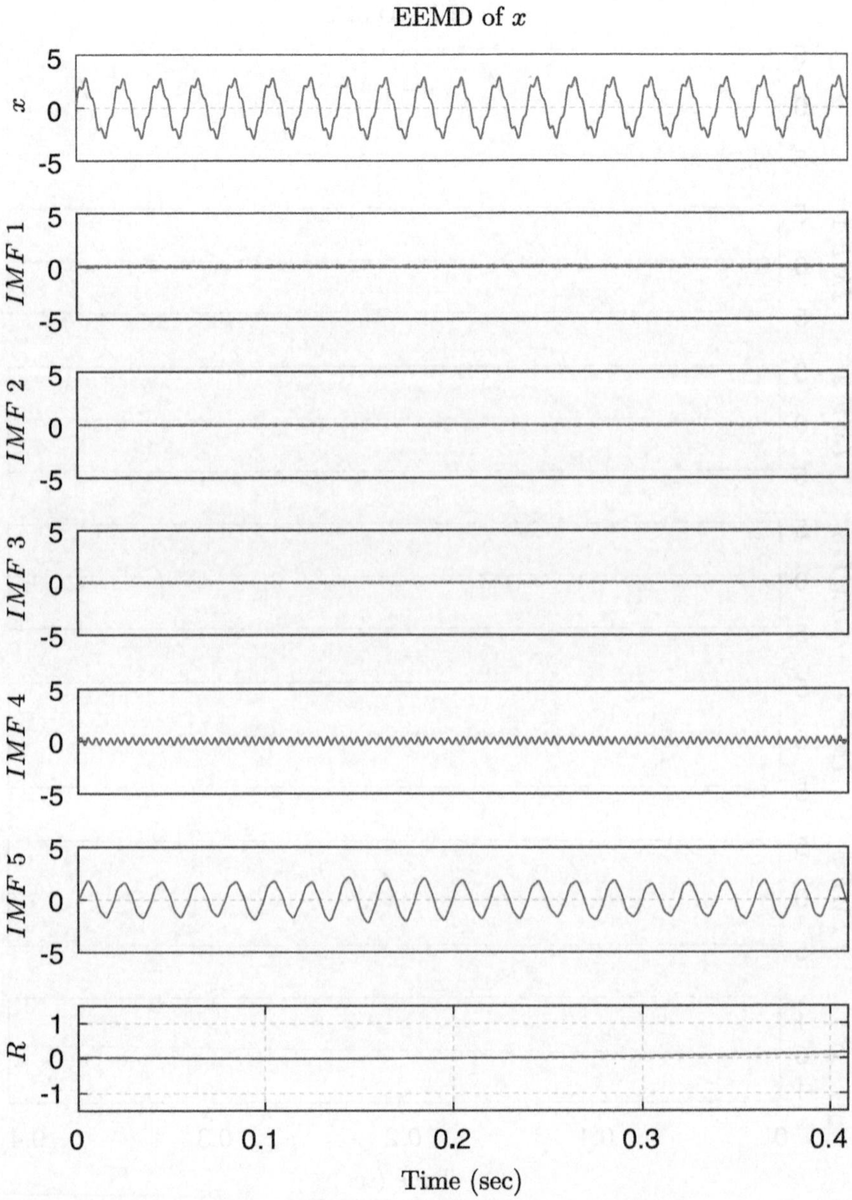

Figure 4.21. *EEMD decomposition of the stator current: Ball defect case*

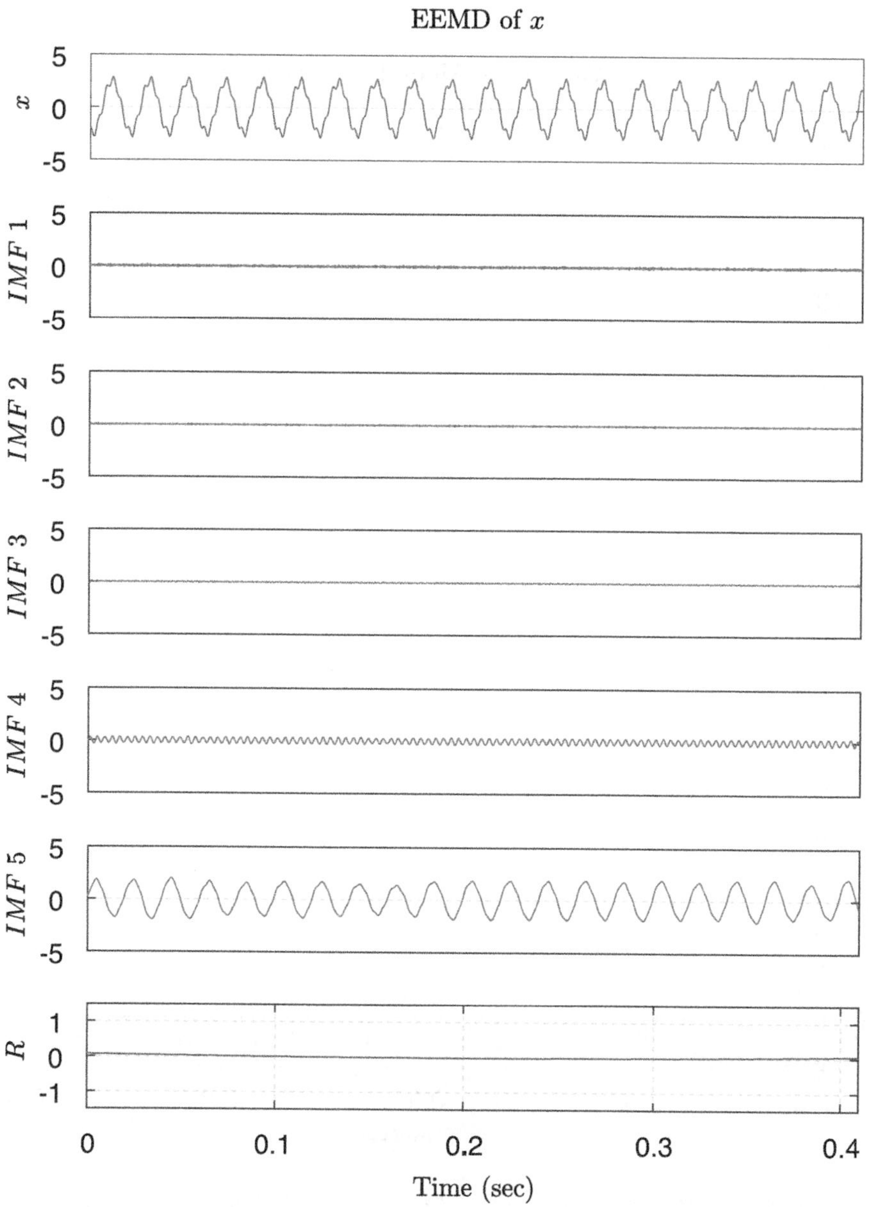

Figure 4.22. *EEMD decomposition of the stator current: Inner race defect case*

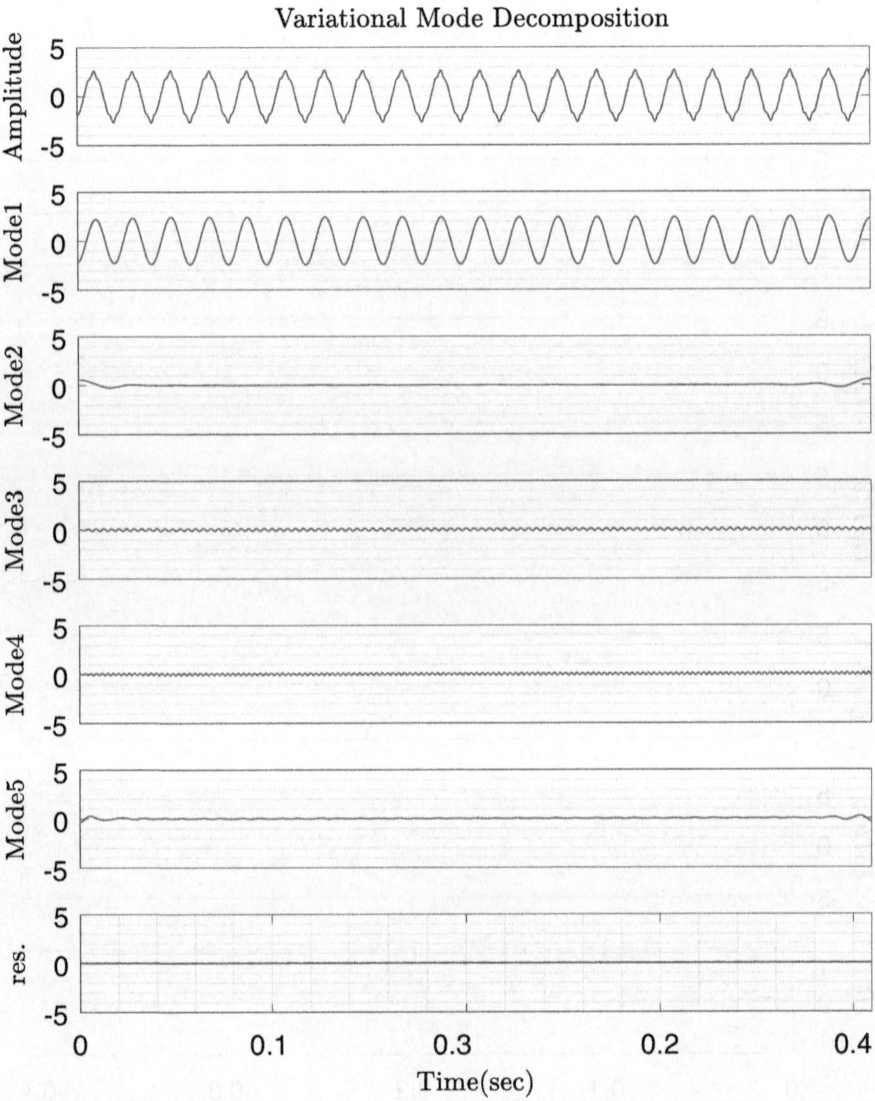

Figure 4.23. *VMD decomposition of the stator current: Healthy bearing case*

It consists of a 0.75 kW electric motor driving a shaft on which a DC generator is mounted. The faults were seeded separately on the rolling elements and on the inner and outer races. Each faulty bearing was reinstalled (separately) on the drive side of the motor shaft, which was then run at a constant speed. The tested induction motor has the following rated parameters: 0.75 kW, 220/380 V, 1.95/3.4 A, 2780 rpm, 50 Hz, 2 poles, Y-connected. It has two 6204.2Z-type bearings.

Stator phase currents are depicted in Figure 4.18 for both healthy and faulty bearings. The EEMD decomposition results are illustrated in Figures 4.19–4.22. Figures 4.23–4.26 displays the set of modes obtained after variational mode decomposition of the analyzed signal $x(n)$ for the healthy case, cage defect case, ball defect case and inner race defect case, respectively.

In order to identify the dominant IMF and Mode, the Pearson correlation coefficient is computed, and the Pearson correlation scores are illustrated in Figures 4.27 and 4.28 for different bearing defects. Pearson correlation scores show clearly that for EEMD the fifth IMF is the closest one, whereas for VMD the first Mode is the dominant one regardless of the type of the fault for both algorithms. Then $\mathrm{IMF_d}$ and $\mathrm{Mode_d}$ are removed from the original signal. In this context, the subtraction process results, after $\mathrm{IMF_d}$ and $\mathrm{Mode_d}$ cancellation, are given in Figures 4.29 and 4.30, where the remaining and useful signals for fault detection are displayed.

Figures 4.27 and 4.28 illustrate the Pearson correlation coefficients for both EEMD and VMD algorithms. This shows that both algorithms are able to extract the dominant mode (i.e., IMF), but not at the same rank; for EEMD, the fifth IMF is the dominant mode, whereas for the VMD, it is the first one. After the dominant mode subtraction, the fault detector σ^2 is evaluated for different cases, and the results are depicted in Figure 4.31. It is noted that σ^2 is not null even in the case of healthy machines for both algorithms. However, it can be noticed that σ^2 for the faulty case is greater than the one under healthy condition. Under faulty condition, σ^2 is five times higher than σ^2 for the healthy case for EEMD algorithm and is multiplied by 65 for the VMD algorithm. The ratio $\sigma^2{}_\mathrm{f}/\sigma^2_\mathrm{h}$ is reported in Table 4.5 for both algorithms and for each fault.

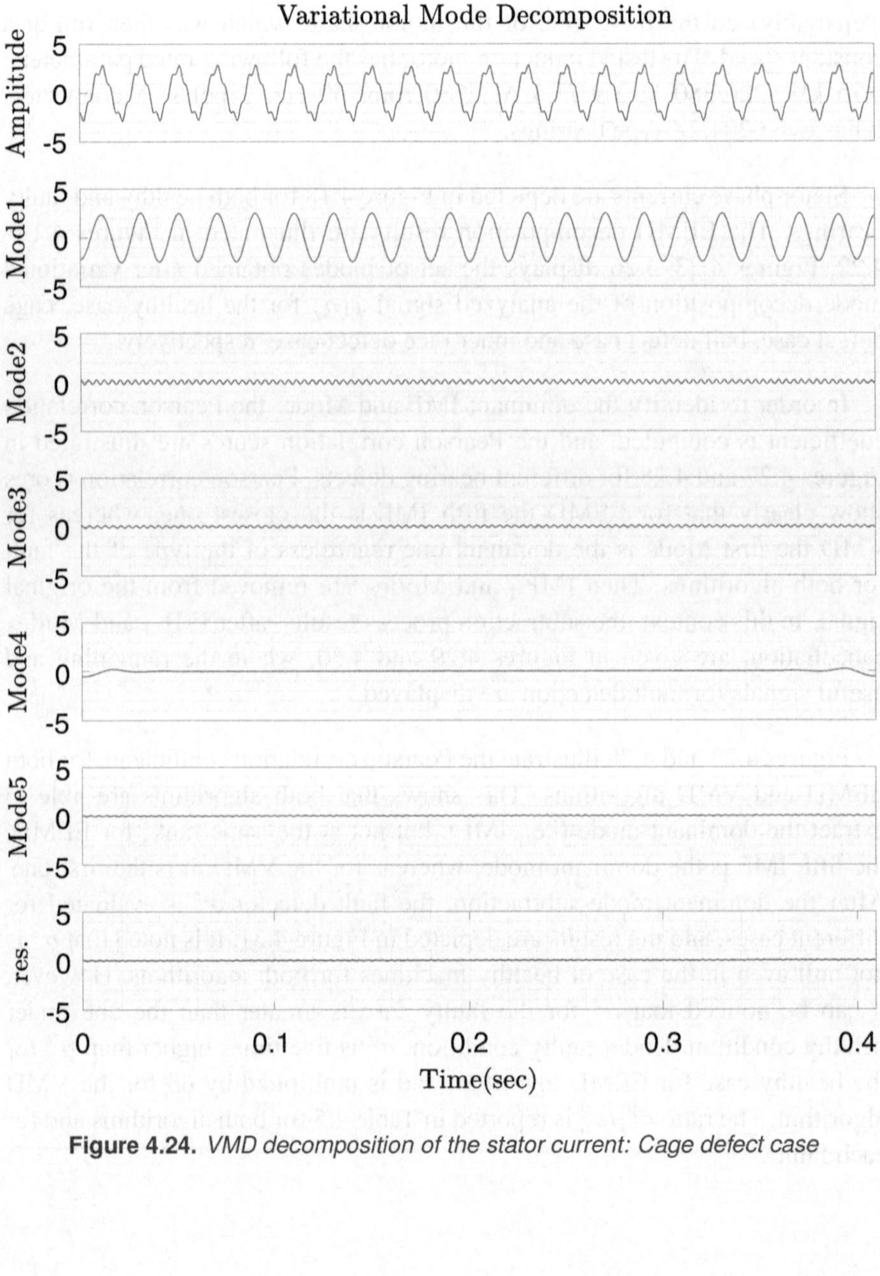

Figure 4.24. *VMD decomposition of the stator current: Cage defect case*

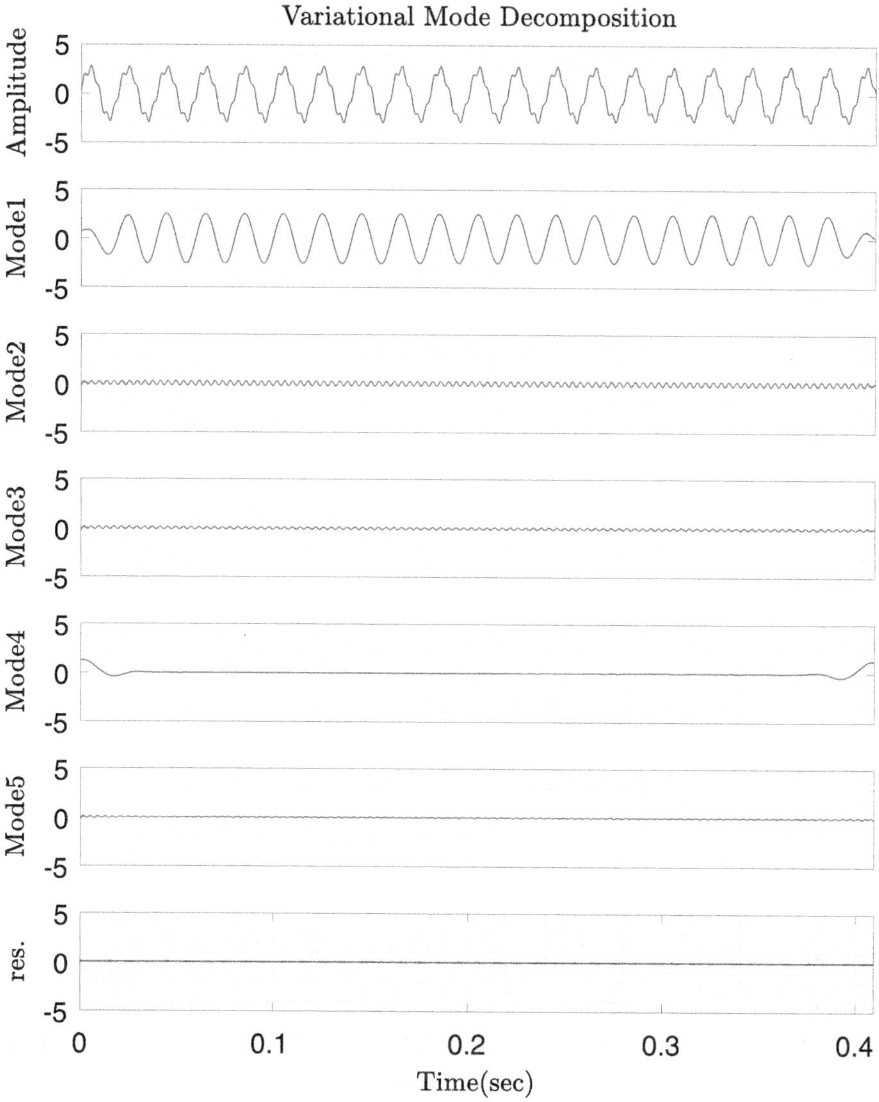

Figure 4.25. *VMD decomposition of the stator current: Ball defect case*

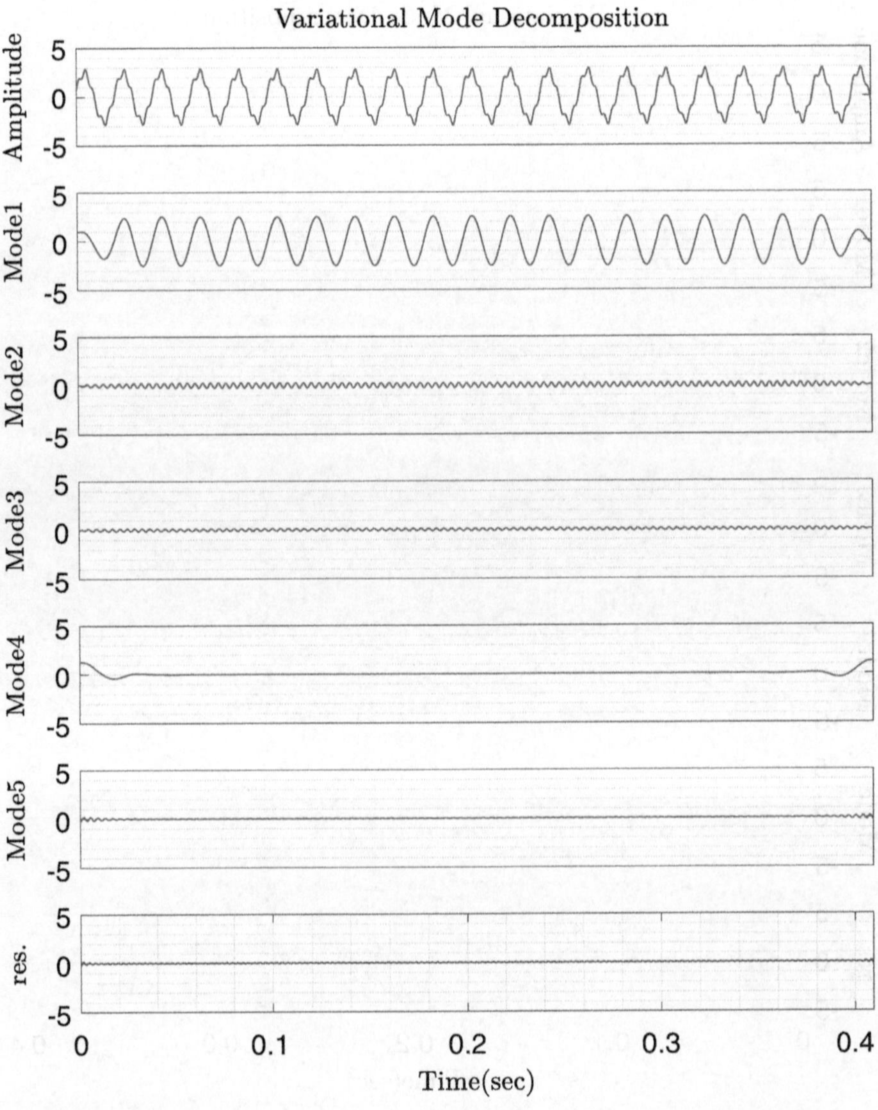

Figure 4.26. *VMD decomposition of the stator current: Inner race defect case*

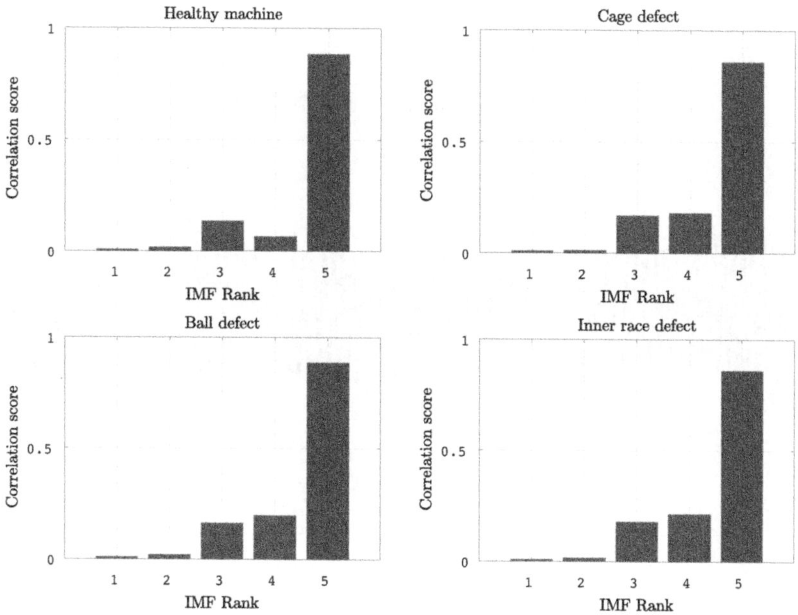

Figure 4.27. *Pearson correlation coefficients for EEMD*

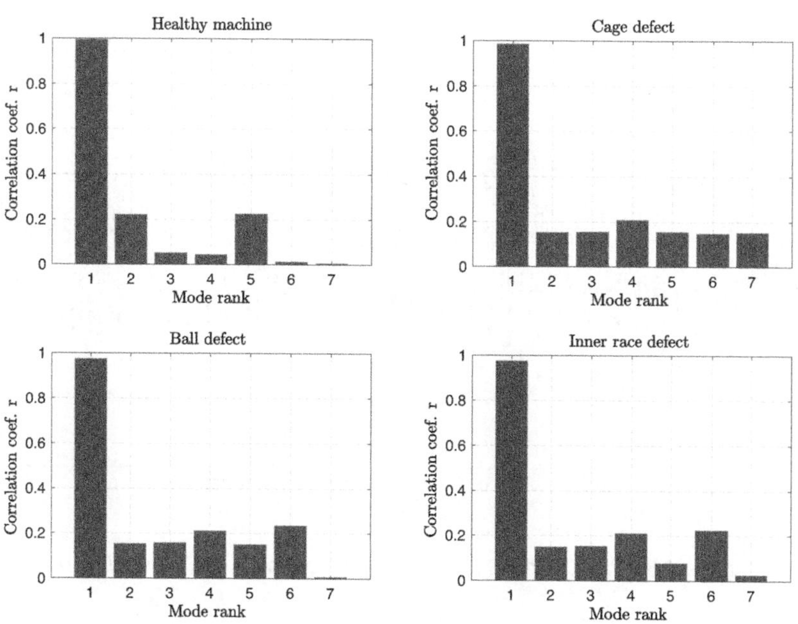

Figure 4.28. *Pearson correlation coefficients for VMD*

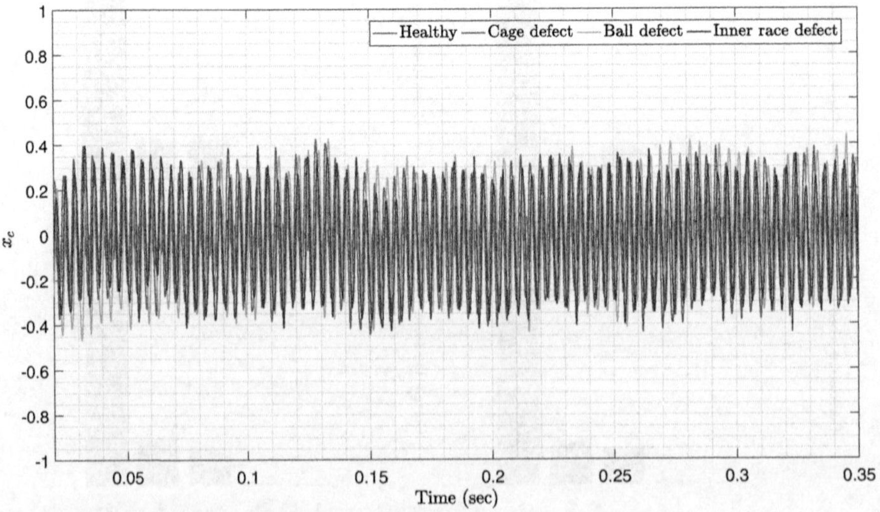

Figure 4.29. *EEMD remain signal x_c after dominant IMF cancellation*

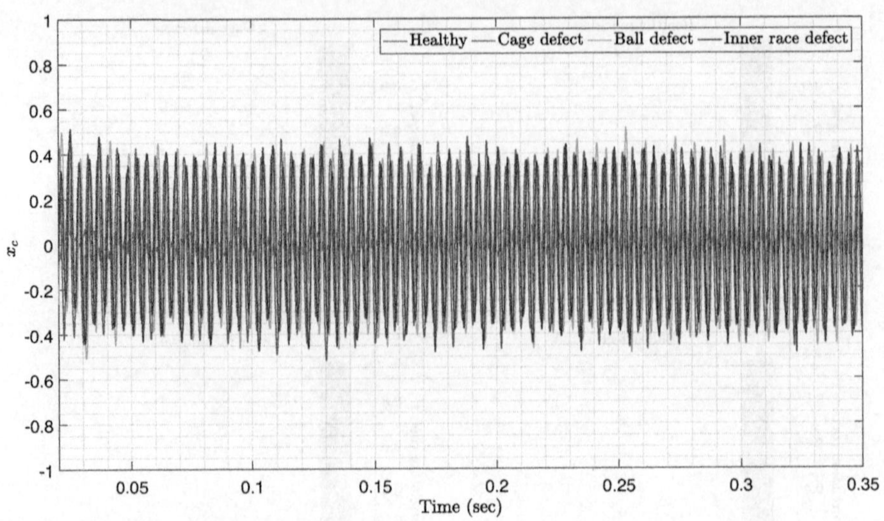

Figure 4.30. *VMD remain signal x_c after dominant IMF cancellation*

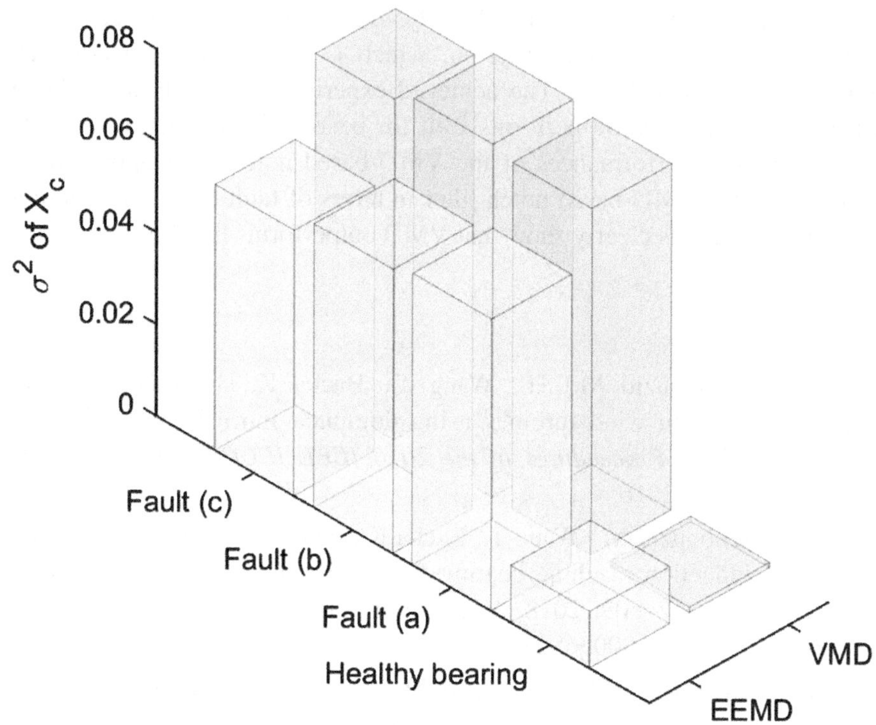

Figure 4.31. *Fault criterion σ^2 for both EEMD and VMD*

Algorithm	Cage defect	Ball defect	Inner race defect
EEMD	5.131	4.997	4.676
VMD	65.492	66.740	64.415

Table 4.5. *Ratio of σ^2 for the faulty case to σ^2 for the healthy case*

4.7. Conclusion

In this chapter, we have proposed an approach based on a low-complexity fault detector for induction machines. This approach is based on a mode decomposition-based notch filter, the core of which is a data-driven strategy combined to a statistical tool. The filtering operation was carried out as follows. The first step concerns the decomposition of the phase machine current into intrinsic mode functions using EEMD or variational mode functions using the VMD algorithm. In the second step, the dominant mode is subtracted from the original signal. Finally, in the last step, a statistical feature is used as a fault detection criterion, which is performed to discriminate between healthy and faulty. The achieved experimental results have shown that the proposed method performs well for bearing faults regardless of the mode rank. The performances of the VMD-based notch filter approach are compared to the EMD-based notch filter in terms of fault detection criterion. The achieved results clearly show that VMD outperforms EEMD.

4.8. References

Amirat Y., Benbouzid M.E.H., Wang T., Bacha K., Feld G., "Bearing fault detection in wind turbines using dominant intrinsic mode function subtraction", in *Proceedings of the 2016 IEEE IECON*, Florence, Italy, pp. 6961–6965, 2016.

Amirat Y., Benbouzid M., Wang T., Bacha K., Feld G., "EEMD-based notch filter for induction machine bearing faults detection", *Applied Acoustics*, vol. 133, pp. 202–209, 2018, available at: http://www.sciencedirect.com/science/article/pii/S0003682X17308125.

Amirat Y., Choqueuse V., Benbouzid M., "Condition monitoring of wind turbines based on amplitude demodulation", in *2010 IEEE Energy Conversion Congress and Exposition*, pp. 2417–2421, 2010.

Amirat Y., Choqueuse V., Benbouzid M.E.H., "EEMD-based wind turbine bearing failure detection using the generator stator current homopolar component", *Mechanical Systems and Signal Processing*, vol. 41, no. 1, pp. 667–678, 2013.

Amirat Y., Elbouchikhi E., Zhou Z., Benbouzid M.E.H., Feld G., "Variational mode decomposition based notch filter for bearing faults detection", in *Proceedings of the 2019 IEEE IECON*, Lisbon, Portugal, pp. 1–6, 2019.

Antonino-Daviu J.A., Riera-Guasp M., Pineda-Sanchez M., Prez R.B., "A critical comparison between DWT and Hilbert–Huang-based methods for the diagnosis of rotor bar failures in induction machines", *IEEE Transactions on Industry Applications*, vol. 45, no. 5, pp. 1794–1803, 2009.

Antonino-Daviu J., Corral-Hernandez J., Climente-Alarcò V., Razik H., "Case stories of advanced rotor assessment in field motors operated with soft-starters and frequency converters", in *IECON 2015 – 41st Annual Conference of the IEEE Industrial Electronics Society*, pp. 001139–001144, 2015.

Benbouzid M.E.H., "A review of induction motors signature analysis as a medium for faults detection", *IEEE Transactions on Industrial Electronics*, vol. 47, no. 5, pp. 984–993, 2000.

Blodt M., Bonacci D., Regnier J., Chabert M., Faucher J., "On-line monitoring of mechanical faults in variable-speed induction motor drives using the Wigner distribution", *IEEE Transactions on Industry Applications*, vol. 55, no. 2, pp. 522–533, 2008.

Blodt M., Regnier J., Faucher J., "Distinguishing load torque oscillations and eccentricity faults in induction motors using stator current Wigner distributions", *IEEE Transactions on Industry Applications*, vol. 45, no. 6, 1991–2000, 2009.

Cardoso A.J.M., *Diagnosis and fault tolerance of electrical machines, power electronics and drives*, IET Energy Engineering, The Institution of Engineering and Technology, UK, 2018.

Cardoso A.J.M., Saraiva S., "Computer-aided detection of airgap eccentricity in operating three-phase induction motors by Park's vector approach", *IEEE Transactions on Industry Applications*, vol. 28, no. 5, pp. 897–901, 1993.

Cardoso A.J.M., Cruz S.M.A., Fonseca D.S.B., "Inter-turn stator winding fault diagnosis in three-phase induction motors, by Park's vector approach", *IEEE Transactions on Energy Conversion*, vol. 14, no. 3, pp. 595–598, 1999.

Chen X., Yang Y., Cui Z., Shen J., "Vibration fault diagnosis of wind turbines based on variational mode decomposition and energy entropy", *Energy*, vol. 174, pp. 1100–1109, 2019, available at: http://www.sciencedirect.com/science/article/pii/S036054421930461X

Choqueuse V., Benbouzid M.E.H., Amirat Y., Turri S., "Diagnosis of three-phase electrical machines using multidimensional demodulation techniques", *IEEE Transactions on Industrial Electronics*, vol. 59, no. 4, pp. 2014–2023, 2012.

Cruz S.M.A., Cardoso A.J.M., "Stator winding fault diagnosis in three-phase synchronous and asynchronous motors, by the extended Park's vector approach", *IEEE Transactions on Industry Applications*, vol. 37, no. 5, pp. 1227–1233, 2001.

Cusido J.C., Romeral L., Ortega J.A., Rosero J.A., Espinosa A.G., "Fault detection in induction machines using power spectral density in wavelet decomposition", *IEEE Transactions on Industrial Electronics*, vol. 55, no. 2, pp. 633–643, 2008.

Dragomiretskiy K., Zosso D., "Variational mode decomposition", *IEEE Transactions on Signal Processing*, vol. 62, no. 3, pp. 531–544, 2014.

Elbouchikhi E., Choqueuse V., Benbouzid M.E.H., "Condition monitoring of induction motors based on stator currents demodulation", *International Review on Electrical Engineering*, vol. 10, no. 6, pp. 1–6, 2015.

Elbouchikhi E., Choqueuse V., Benbouzid M.E.H., "Induction machine faults detection using stator current parametric spectral estimation", *Mechanical Systems and Signal Processing*, vol. 5253, pp. 447–464, 2015.

Elbouchikhi E., Choqueuse V., Benbouzid M., "Induction machine bearing faults detection based on a multi-dimensional music algorithm and maximum likelihood estimation", *ISA Transactions*, vol. 63, pp. 413–424, 2016.

Frosini L., Harlişca C., Szabó L., "Stator current and motor efficiency as indicators for different types of bearing fault in induction motor", *IEEE Transactions on Industrial Electronics*, vol. 57, no. 1, pp. 244–251, 2010.

Frosini L., Harlişca C., Szabó L., "Induction machine bearing fault detection by means of statistical processing of the stray flux measurement", *IEEE Transactions on Industrial Electronics*, vol. 62, no. 3, pp. 1846–1854, 2015.

Gilles J., "Empirical wavelet transform", *IEEE Transactions on Signal Processing*, vol. 61, no. 16, pp. 3999–4010, 2013.

Guo L., Chen J., Li X., "Rolling bearing fault classification based on envelope spectrum and support vector machine", *Journal of Vibration and Control*, vol. 15, no. 9, pp. 1349–1363, 2009.

Harmouche J., Delpha C., Diallo D., "Improved fault diagnosis of ball bearings based on the global spectrum of vibration signals", *IEEE Transactions on Energy Conversion*, vol. 30, no. 1, pp. 376–383, 2015.

He K., Tso G.K., Zou Y., Liu J., "Crude oil risk forecasting: New evidence from multiscale analysis approach", *Energy Economics*, vol. 76, pp. 574–583, 2018.

Huang N.E., Shen S.S.P., *Hilbert–Huang transform and its applications*, Interdisciplinary Mathematical Sciences, Singapore, 2014.

Huang N.E., Shen Z., Long S.R., Wu M.L., Shih H.H., Zheng Q., Yen N.C., Tung C.C., Liu H.H., "The empirical mode decomposition and Hilbert spectrum for nonlinear and nonstationary time series analysis", *Proceedings of Royal Society, London*, vol. 454, pp. 903–995, 1998.

Ibrahim A., Badaoui M.E., Guillet F., Bonnardot F., "A new bearing fault detection method in induction machines based on instantaneous power factor", *IEEE Transactions on Industrial Electronics*, vol. 55, no. 12, pp. 4252–4259, 2008.

Jiang F., Zhu Z., Li W., "An improved VMD with empirical mode decomposition and its application in incipient fault detection of rolling bearing", *IEEE Access*, vol. 6, pp. 44483–44493, 2018.

Kamwa I., Pradhan A.K., Joos G., "Robust detection and analysis of power system oscillations using the Teager–Kaiser energy operator", *IEEE Transactions on Power Systems*, vol. 26, no. 1, pp. 323–333, 2011.

Kia S.H., Henao H., Capolino G.A., "A high-resolution frequency estimation method for three-phase induction machine fault detection", *IEEE Transactions on Industrial Electronics*, vol. 54, no. 4, pp. 2305–2314, 2007.

Kia S.H., Henao H., Capolino G.A., "Diagnosis of broken-bar fault in induction machines using discrete wavelet transform without slip estimation", *IEEE Transactions on Industry Applications*, vol. 45, no. 4, pp. 1395–1404, 2009.

Kliman G.B., Koegl R.A., Stein J., Endicott R.D., Madden M.W., "Noninvasive detection of broken rotor bars in operating induction motors", *IEEE Transactions on Energy Conversion*, vol. 3, no. 4, pp. 873–879, 1988.

Komaty A., Boudraa A.O., Augier B., Dar-Emzivat D., "EMD-based filtering using similarity measure between probability density functions of IMFs", *IEEE Transactions on Instrumentation and Measurement*, vol. 63, no. 1, pp. 27–34, 2014.

Kusiak A., "Renewables: Share data on wind energy", *Nature*, vol. 522, no. 1, pp. 19–21, 2016.

Leite V.C.M.N., da Silva J.G.B., Veloso G.F.C., da Silva L.E.B., Lambert-Torres G., Bonaldi E.L., d. L. de Oliveira, L.E., "Detection of localized bearing faults in induction machines by spectral kurtosis and envelope analysis of stator current", *IEEE Transactions on Industrial Electronics*, vol. 62, no. 3, pp. 1855–1865, 2015.

Mandic D.P., Ur Rehman N., Wu Z., Huang N.E., "Empirical mode decomposition-based time–frequency analysis of multivariate signals: The power of adaptive data analysis", *IEEE Signal Processing Magazine*, 30(6), pp. 74–86, 2013.

Miao Q., Makis V., "Condition monitoring and classification of rotating machinery using wavelets and hidden Markov models", *Mechanical Systems and Signal Processing*, vol. 21, no. 2, pp. 840–855, 2007.

Nejjari H., Benbouzid M.E.H., "Monitoring and diagnosis of induction motors electrical faults using a current Park's vector pattern learning approach", *IEEE Transactions on Industry Applications*, vol. 36, no. 3, pp. 730–735, 2000.

Ocak H., Loparo K.A., "A new bearing fault detection and diagnosis schema based on hidden Markov modeling of vibration signals", in *Proceedings of IEEE International Conference on Acoustics, Speech, and Signal Processing ICASSP*, Salt Lake City, USA, pp. 3141–3144, 2001.

Onel I.Y., Benbouzid M.E.H., "Induction motor bearing failure detection and diagnosis: Park and concordia transform approaches comparative study", *IEEE/ASME Transactions on Mechatronics*, vol. 13, no. 2, pp. 257–262, 2008.

Phadke A.G., "Synchronized phasor measurements – A historical overview", in *IEEE/PES Transmission and Distribution Conference and Exhibition*, vol. 1, pp. 476–479, 2002.

Proakis J.G., Manolakis D.G., *Digital signal processing*, 3rd ed., Prentice Hall, New Jersey, 1996.

Riera-Guasp M., Antonio-Daviu J.A., Roger-Folch J., Palomares M.P.M., "The use of the wavelet approximation signal as a tool for the diagnosis of rotor bar failure", *IEEE Transactions on Industry Applications*, vol. 44, no. 3, pp. 716–726, 2008.

Riley C.M., Lin B.K., Habetler T.G., Schoen R.R., "A method for sensorless on-line vibration monitoring of induction machines", *IEEE Transactions on Industry Applications*, vol. 34, no. 6, pp. 1240–1245, 1998.

Schoen R.R., Habetler T.G., Kamran F., Bartfield R.G., "Motor bearing damage detection using stator current monitoring", *IEEE Transactions on Industry Applications*, vol. 31, no. 6, pp. 1274–1279, 1995.

Seera M., Lim C.P., "Online motor fault detection and diagnosis using a hybrid FMM-cart model", *IEEE Transactions on Neural Networks and Learning Systems*, vol. 25, no. 4, pp. 806–812, 2014.

Stack J.R., Harley R.G., Habetler T.G., "An amplitude modulation detector for fault diagnosis in rolling element bearings", *IEEE Transactions on Industrial Electronics*, vol. 51, no. 5, pp. 1097–1102, 2004.

Stoica P., Moses R., *Introduction to spectral analysis*, Prentice Hall, New Jersey, 1997.

Tanaka T., Mandic D.P., "Complex empirical mode decomposition", *IEEE Letters on Signal Processing*, vol. 14, no. 2, pp. 101–104, 2007.

Tavner P.J., "Review of condition monitoring of rotating electrical machines", *IET Electric Power Applications*, vol. 2, no. 4, pp. 215–247, 2008.

Tavner P., Ran L., Penman J., Sedding H., *Condition monitoring of rotating electrical machines*, 2nd ed., IET, London, 2008.

Thomson W.T., Stewart I.D., "Online current monitoring for fault diagnosis in inverter fed induction motors", in *Third International Conference on Power Electronics and Variable-Speed Drives*, pp. 432–435, 1988.

Thorsen O.V., Dalva M., "Failure identification and analysis for high-voltage induction motors in the petrochemical industry", *IEEE Transactions on Industry Applications*, vol. 35, no. 4, pp. 810–818, 1999.

Torres M.E., Colominas M., Schlotthauer G., Flandrin P., "A complete ensemble empirical mode decomposition with adaptive noise", in *Proceedings of 2011 IEEE International Conference on Acoustics, Speech and Signal Processing (ICASSP)*, pp. 4144–4147, 2011.

Trajin B., Chabert M., Regnier J., Faucher J., "Hilbert versus concordia transform for three phase machine stator current time–frequency monitoring", *Mechanical Systems & Signal Processing*, vol. 23, no. 8, pp. 2648–2657, 2009.

Wang W., Jianu, O.A., "A smart sensing unit for vibration measurement and monitoring", *IEEE/ASME Transactions on Mechatronics*, vol. 15, no. 1, pp. 70–78, 2010.

Wang Z., He G., Du W., Zhou J., Han X., Wang J., He H., Guo X., Wang J., Kou Y., "Crude oil risk forecasting: New evidence from multiscale analysis approach", *IEEE Access*, vol. 7, pp. 44871–44882, 2019.

Wu Z.H., Huang N.E., "Ensemble empirical mode decomposition: A noise-assisted data analysis method", *Advances in Adaptive Data Analysis*, vol. 1, pp. 1–41, 2009.

Yang W., Tavner P.J., Crabtree C.J., Wilkinson M., "Cost-effective condition monitoring for wind turbines", *IEEE Transactions on Industrial Electronics*, vol. 57, no. 1, pp. 263–271, 2010.

Yu D., Cheng J., Yang Y., "Application of EMD method and Hilbert spectrum to the fault diagnosis of roller bearings", *Mechanical Systems and Signal Processing*, vol. 19, no. 2, pp. 259–270, 2005.

Zhang P., Du Y., Habetler T., Lu B., "A survey of condition monitoring and protection methods for medium-voltage induction motors", *IEEE Transactions on Industry Applications*, vol. 47, no. 1, pp. 34–46, 2011.

Zhao H., Li L., "Fault diagnosis of wind turbine bearing based on variational mode decomposition and Teager energy operator", *IET Renewable Power Generation*, vol. 11, no. 4, pp. 453–460, 2017.

Zho W., Habetler T., Harley R.G., "Bearing fault detection via stator current noise cancellation and statistical control", *IEEE Transactions on Industrial Electronics*, vol. 55, no. 12, pp. 4260–4269, 2008.

5

Methods for Lifespan Modeling in Electrical Engineering

5.1. Introduction

This chapter presents and compares statistical methods for electrical device lifespan modeling from small-sized training sets. Nowadays, an increasing number of applications use electrical power. In particular, the aircraft industry moves toward "more electrical aircrafts" by replacing heavy conventional mechanical, hydraulic or pneumatic systems by electrical ones [FAN 13]. More electrical aircrafts aim at reducing weight, complexity, fuel consumption, operational costs and environmental impact and, at the same time, demand high reliability and high availability. However, increase in electrical aircraft put more challenges on the electrical systems. Particularly, the increased power demand leads to a growing risk of partial discharge (PD) in the insulation systems [CHR 11]. Unfortunately, a large number of electrical failures is related to insulation materials [TAV 08]. Hence, this chapter mainly considers element lifespan modeling. Considering the average lifespan (several thousand hours under nominal stress for most electrical devices), carrying out full aging tests is expensive. Consequently, the devices are exposed to overstress in order to bring the lifespan under an acceptable level [GUA 11]. The objective of the proposed methods is twofold: the prediction of lifespan under nominal conditions as well as a better understanding of the coupling effects between different stress factors. Admittedly, there exist some physical models that relate the insulation material lifespan to a given stress factor or even to several factors [MAZ 09, STO 93]. Unfortunately, these models require measurement of some

---

Chapter written by Antoine PICOT, Marie CHABERT and Pascal MAUSSION.

physical constants through generally cumbersome experimental setups. Moreover, they do not provide a comprehensive description of the stress factor interactions. Because the proposed methods are not specific for insulation material lifespan modeling, they are also applied to organic light-emitting devices (OLEDs).

The statistical methods for lifespan modeling presented in this chapter can be split into two classes: parametric and nonparametric. Parametric methods describe the effects of factors and interactions through a mathematical relation. This relationship involves parameters that must be estimated from the learning set. Statistical tests then enable us to assess the significance of each parameter. As an alternative, nonparametric methods such as regression trees classify experiments in different operational modes for which the factors are ranked as a function of their effect on the lifespan. Then, the lifespan can be represented graphically as a function of the stress factors. Unfortunately, regression trees provide a piecewise constant model of the lifespan and thus a less fine modeling than the parametric models. Finally, hybrid models, which combine the advantages of both parametric and nonparametric models, are considered. They estimate a parametric model in each class identified by the regression tree. The resulting model is piecewise multilinear and thus a finer analysis of the influence of several stress factors.

Among parametric methods, the two first ones, the design of experiments (DoE) [FIS 35, LAH 13] and the response surface method (RSM) [MYE 02, PIC 13], take into account experimental constraints. For its higher flexibility with respect to the choice of the predictor variables and to their ranges of variation, multiple linear regression (MLR) [SAL 15] is also studied. Earlier, the significance of different factors is studied using analysis of variance (ANOVA) [PIC 13]. Finally, the bootstrap method is introduced to assess the validity of the obtained lifespan parametric models in the case of small-size data sets [EFR 93]. This method also enables the validation of some underlying assumptions on the lifespan model.

5.2. Parametric methods

5.2.1. *Design of experiments*

A basic method for estimating the parameters of a model consists of varying the variables one-by-one. This method, called one factor at a time

(OFAT), is time-consuming and also requires a lot of effort. Moreover, it is accurate only when there is no coupling effect between the factors. The DoE is a systematic and efficient method that proposes a model and determines the best combination of factor levels for an optimized experiment design, in order to identify its coefficients. It has the following advantages:

– the optimal solution is obtained using less number of experiments;

– increased model accuracy, assuming an isovariance for each of the n experiments, the variance for each parameter estimate will be divided by $n^{1/2}$;

– several factors are considered in the same experimental plan (instead of OFAT) and interactions between factors can be taken into account;

– the significance of the factor is assessed through a statistical hypothesis test.

Experimental design started during the 1930s with Fisher's work [FIS 35] and has been developed significantly after the publication of some predefined tables by Taguchi [TAG 87]. The objective is to design a schedule of experiments to obtain the largest possible amount of information with a minimum number of experiments. This method demonstrated its performance in different application domains, especially in chemistry and mechanics, where a high volume of factors have to be tuned simultaneously. The method has already proved its effectiveness in electrical engineering [ZER 13, ZHU 14], including the insulation lifespan modeling [LAH 13, SAL 17].

Experiment no.	M	Factor 1	Factor 2	Interaction 12	Criterion Y
1	1	−1	−1	+1	y_1
2	1	+1	−1	−1	y_2
3	1	−1	+1	−1	y_3
4	1	+1	+1	+1	y_4
					M

Table 5.1. *Example of an experimental table*

In full factorial designs, a set of experiments is designed by systematically varying all the relevant factors. The number of values for each factor is set *a priori* and defines the different factor levels. The main purpose is to obtain information on the effects of the selected parameters. The idea is to modify the level of each factor in each experiment according to a specific procedure. This allows for a drastic reduction in the number of required experiments. A simple example with only two parameters with two levels each is presented in Table 5.1. Normalized (centered reduced) factors are used, that is, –1 for the low level and +1 for the high level of each factor.

Each row represents an experiment and each column is indexed to a factor. For each experiment i, the selected criterion value y_i, which is the system response, is measured. In the DoE, the effect of a factor is obtained according to expression [5.1]. For example, $E_1 = 0.12$ means that factor 1 at its highest level has an effect of +0.12 on the system response, normalized with respect to its average value dented by M. The main advantage of the DoE is that the effect of interactions between factors can also be determined. Equation [5.2] presents the average effect of interaction E_{12} between factors 1 and 2 on the desired criterion:

$$E_1 = \frac{-y_1 + y_2 - y_3 + y_4}{4}, \qquad [5.1]$$

$$E_{12} = \frac{y_1 - y_2 - y_3 + y_4}{4}. \qquad [5.2]$$

The mathematical model relating response Y to different factors is

$$Y \sim M + E_1 F_1 + E_2 F_2 + E_{12} I_{F1F2}, \qquad [5.3]$$

where Y is the system response vector (the insulation lifespan in the case of study), M is the mean value of the system responses, F_i are the factor levels (normalized thermal and electrical stresses imposed in the aging test), I_{FiFj} are the levels of the interactions between factors F_i and F_j, E_i are the effects of the factors and E_{ij} are the effects of the interactions between factors F_i and F_j on the global response.

Furthermore, it is easy to verify that the effect vector $E = [M \quad E_1 \quad E_2 \quad E_{12}]^{\mathrm{T}}$ can be calculated through

$$E = X^{-1} Y, \qquad [5.4]$$

where X is the matrix obtained from the experimental plan as shown in the gray part of Table 5.1 and Y is the vector of the experimental results presented in the last column.

5.2.2. *Response surfaces*

DoE provides an appropriate model estimation only if the hypothesis of a multilinear relationship between the response, the factors and their interactions is valid. However, such a model is not always sufficient. The RSM model is the extension of the DoE model, which considers the quadratic effects of the main factors [MYE 02, BRA 07, PIC 13]. Consequently, k additional terms are added to the model (where k denotes the number of factors). Similar to DoE, RSM also organizes the experiments and minimizes their number, while providing a more complex and thus finer model at the same time with additional experiments: some central experiments (all levels equal to 0 when considering centered reduced variables) must also be tested. Consequently, the RSM design requires at least three levels. Fortunately, the experiments performed for DoE can also be used for RSM.

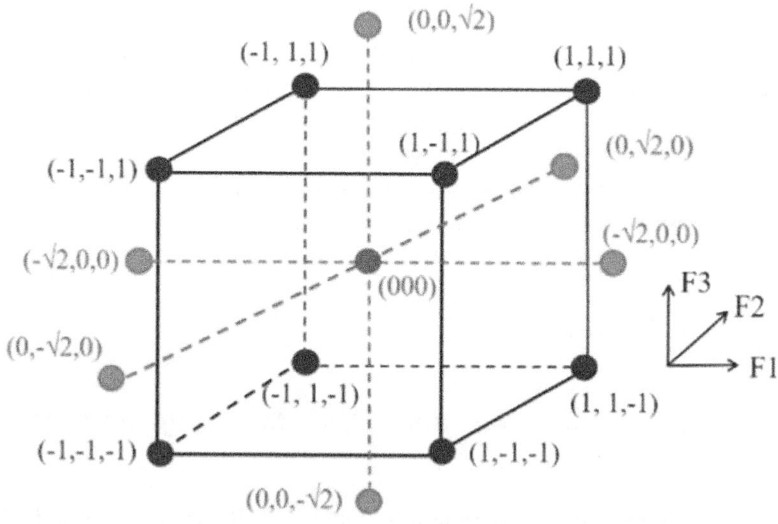

Figure 5.1. *Graphical representation of the experimental points in a DoE (black) and in a CCD (in red) with $\mu = \pm\sqrt{2}$ for three factors (two levels each in a DoE and five levels each in an RSM). For a color version of the figures in this book, see www.iste.co.uk/soualhi/electrical1.zip*

The central composite design (CCD) is the most popular design of RSM, which considers an extra level $\mu = \pm\sqrt{2}$ for several chosen factors and is designed as

– a complete $n_f = 2^k$ DoE design;

– n_0 repetitions at the design center;

– two points on the axis of each parameter situated at a distance μ from the center.

The total number of required experiments in RSM is therefore $n = n_f + n_0 + 2k$.

The values of μ and n_0 are chosen to achieve some particular properties such as orthogonality, rotability, uniform precision, optimality criteria, and so on [MYE 02]. Orthogonal designs are of particular interest as they lead to uncorrelated model coefficients with minimum variances, thus improving the accuracy of the obtained model. The design is orthogonal if μ and n_0 are chosen such that

$$(n_f + 2\mu^2)^2 = n_f n. \qquad [5.5]$$

To guarantee the orthogonality with four experiments in the center of the experimental domain, and according to [5.5], μ must be equal to $\sqrt{2}$, as shown in Figure 5.1. Consequently, five levels $[-\sqrt{2}; -1; 0; +1; +\sqrt{2}]$ are chosen for each factor. The effect vector $\hat{E}$ can be computed in a manner similar to classical DoE methodology, but since X matrix is no longer square (the number of experiments is larger than the number of variables), $\boldsymbol{E}$ is estimated by ordinary least-square (OLS) according to

$$E = (X'X)^{-1}X'Y. \qquad [5.6]$$

5.2.3. *Application of DoE and RSM to aging and lifespan modeling*

In electrical insulation degradation, DoE and RSM have already been successfully used on kapton films and on different twisted pairs under thermal and electrical stress [LAH 13, SAL 15, SAL 17]. DoE has been used to model the lifespan and the luminance degradation of OLEDs under thermal and electrical combined stress [SAL 19]. In the case of lifespan modeling, the factors are related to the electrical and thermal stresses affecting the insulation degradation, and the system response is the insulation lifespan. These designs include all combinations of every factor level, and the number of experiments depends on the number of factor levels.

5.2.3.1. *Experiments on electrical insulation systems of electrical machines*

Experiments were carried out on twisted pairs composed of wires coated with a double electrical insulator varnish which degrades over time when submitted to electrical and thermal stresses. The insulator is designed to last for several thousand hours but, if the nominal constraints are exceeded, its lifespan decreases dramatically. PDs appear in the insulation of rotating machines fed by inverters under high voltage stress [PER 92, GUA 11, ILL 14, STO 05].

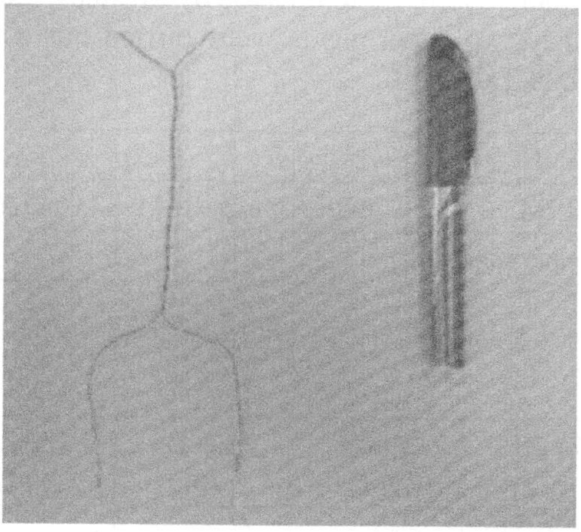

Figure 5.2. *Twisted pairs EDERFIL C200 as test samples, with a double insulating varnish of PEI and PAI*

Tested materials are twisted pairs covered with a double layer of poly-ether-imide (PEI) and poly-amide-imide (PAI) varnish with a thermal class of 200°C (Ederfil C200, diameter 0.5 mm), as shown in Figure 5.2. These twisted pairs are manufactured according to the American National Standards Institute [ANS 07].

The first step consists of defining the stress factors and their range of variation. Experiments are based on accelerated life testing (ALT) [QI 11], that is, materials are submitted to high constraints to test their degradation without waiting until normal failure. Voltage and frequency are the principal causes of PDs into the insulator [MON 89, MON 02, MON 11]. The extreme stress values are chosen either to accelerate the degradation or to represent extreme conditions. They are given in Table 5.2.

Factors	Minimum Value	Maximum Value
Voltage (kV)	1	3
Frequency (kHz)	5	15
Temperature (°C)	–55	180

Table 5.2. *Extreme values of the three stress factors*

Six twisted pairs were simultaneously tested under each stress configuration. Thirty-two experiments were carried out with six repetitions for each, as shown in Table 5.3.

DoE	8 exp.
RSM	DoE + 6 exp. RSM + 4 central points
Extra experiments	12 exp.

Table 5.3. *Number of experiments for the three methods*

Factors	Level ($-\sqrt{2}$)	Level (–1)	Level (0)	Level (+1)	Level ($+\sqrt{2}$)
log (10*Voltage) (kV)	log(10*1)	log(10*1.17)	log(10*1.3)	log(10*2.5)	log(10*3)
log (frequency) (kHz)	log(5)	log(5.87)	log(8.7)	log(12.77)	log(15)
Exp ($-bT$) (°C)	exp(55b)	exp(34.8b)	exp(–26.1b)	exp(–119.7b)	exp(–180b)

Table 5.4. *Normalized levels of the three stress factors for centered and reduced variables*

Some of these experiments are used to estimate the DoE and RSM models and compose the so-called learning set. The other experiments are used to test the validity of these models and compose the so-called test set.

First, the form of variation of the lifespan with respect to each of the three stress factors must be identified. According to Salameh *et al.* [SAL 15], in the case of PEI films under the PD regime, the insulation lifespan logarithm follows an inverse power model depending on log(V), log(F) and exp($-bT$). This model is in accordance with physical laws and also with the ALT method [QI 11].

After defining the stress ranges, forms and levels, the experimental design is now fully specified. Table 5.4 displays the constraint values in each experiment and the corresponding measured lifespan. Experiments 1–8 correspond to DoE and are represented by the black dots in Figure 5.1, whereas the red dots correspond to experiments 9–20 and are used additionally in the RSM model. Finally, experiments 21–32 have a random position with respect to black and red dots and serve as extra experiments either to validate DoE and RSM models or to generate regression models with randomly selected data points. Note that in the DoE and RSM models, the X matrix consists of factor levels, whereas in the regression models, the real values of factors are used instead, as shown in Table 5.5.

In this section, models are estimated according to the DoE and RSM methods. Thanks to DoE, the lifespan model M_1 can be expressed as

$$\begin{aligned}\log(L) \approx\ & M + E_V \cdot \log(10V) + E_F \cdot \log(F) + E_T \cdot \exp(-b \cdot T) \\ & + I_{VF} \cdot \log(10V) \cdot \log(F) + I_{VT} \cdot \log(10V) \cdot \exp(-b \cdot T) \\ & + I_{FT} \cdot \log(F) \cdot \exp(-b \times T) + I_{VFT} \cdot \log(10V) \cdot \log(F) \cdot \exp(-b \cdot T).\end{aligned} \quad [5.7]$$

The first eight experiments in Table 5.5 (learning set) are used to estimate the DoE model coefficients; the model is then tested on the remaining 24 experiments (test set). The estimated values of M, E_V, E_F, …, I_{VFT} are shown in Figure 5.3. The most influential factors and/or interactions are those having the highest effects, that is, the voltage, the temperature and their interaction. The average, maximum and minimum relative errors between the predicted and measured lifespan are shown in Table 5.6. It can be seen from this table that this model does not provide a very accurate lifespan prediction on the 24 extra experiments composing the test set. It is due to the fact that only eight experiments are used to compute this model.

		V (kV)	*F* (kHz)	*T* (°C)	Exp1	Exp2	Exp3	Exp4	Exp5	Exp6
1	DoE	1.174	5.872	–34.82	60	75	81	40.05	55.5	64.5
2	DoE	1.174	5.872	119.74	12.3	14.28	16.0	10.5	14.8	16
3	DoE	1.174	12.77	–34.82	10	19	25.5	21.4	36	44
4	DoE	1.174	12.77	119.74	7	7.66	7.66	4.26	4	4.91
5	DoE	2.554	5.872	–34.82	2.11	4.26	5.45	5.4	5.48	6
6	DoE	2.554	5.872	119.74	1.98	2.316	2.26	1.75	2.25	2.26
7	DoE	2.554	12.77	–34.82	1.05	1.5	1.66	1.3	1.6	1.85
8	DoE	2.554	12.77	119.74	0.21	0.75	0.9	0.2	0.88	0.95
9	RSM	1	8.66	26.12	19.6	20.73	25.2	26.02	26.7	30.23
10	RSM	3	8.66	26.12	0.12	0.27	0.43	0.58	0.7	0.8
11	RSM	1.732	5	26.12	2.17	3.05	5.02	7	8.23	8.55
12	RSM	1.732	15	26.12	1.78	2.18	2.18	2.37	2.65	2.78
13	RSM	1.732	8.66	180	2.36	2.43	2.65	2.44	2.35	2.47
14	RSM	1.732	8.66	–55	13	26.5	28.5	11.5	14.5	16.6
15	RSM	1.732	8.66	26.12	4.35	1.78	5.05	4.62	2.87	5.03
16	Central points	1.732	8.66	26.12	3.35	4.25	6.2	3.92	2.88	3.53
17		1.732	8.66	26.12	5.15	1.7	5.22	4.58	5.03	4.93
18		1.732	8.66	26.12	4.3	5.72	4.65	4.42	4.33	4.77
19		1.732	8.66	26.12	4.17	3.27	4.75	5.05	6.13	4.35
20		1.732	8.66	26.12	4.83	3.82	3.52	4.42	5.72	3.8
21	Randomly selected points	2.5	14.52	–35	0.57	1.3	1.5	1.18	1.27	1.47
22		1.5	10	–15	13.7	16.17	17.6	17.17	18.9	21.1
23		2.81	14	–10	0.32	0.52	0.7	0.33	0.7	0.87
24		2.53	13.79	2	0.55	0.85	1.27	0.98	1	1.28
25		2.56	5.4	10	2.82	3.53	4.28	1.66	4.77	4.92
26		2	10.97	36	2.9	3.55	4.58	2.17	2.85	4.03
27		2.3	7.48	40	1.43	2.27	3.97	3.05	4.08	4.55
28		1.01	11.54	61	12	13.58	17.5	16.5	17.8	19.83
29		1.54	10.58	76	1.83	4.56	4.95	4.33	5.5	5.8
30		1.92	9.68	90	2.5	2.65	3.4	2.8	3.05	3.28
31		1.1	7.98	164	9.75	9.91	10.1	9.2	9.3	9.65
32		1.31	7.79	180	5.5	5.83	5.95	4.9	5	5.43

Table 5.5. *Values of the stress constraints in the 32 experiments for DoE, RSM and MLR and of the six corresponding measured lifespans (in minutes)*

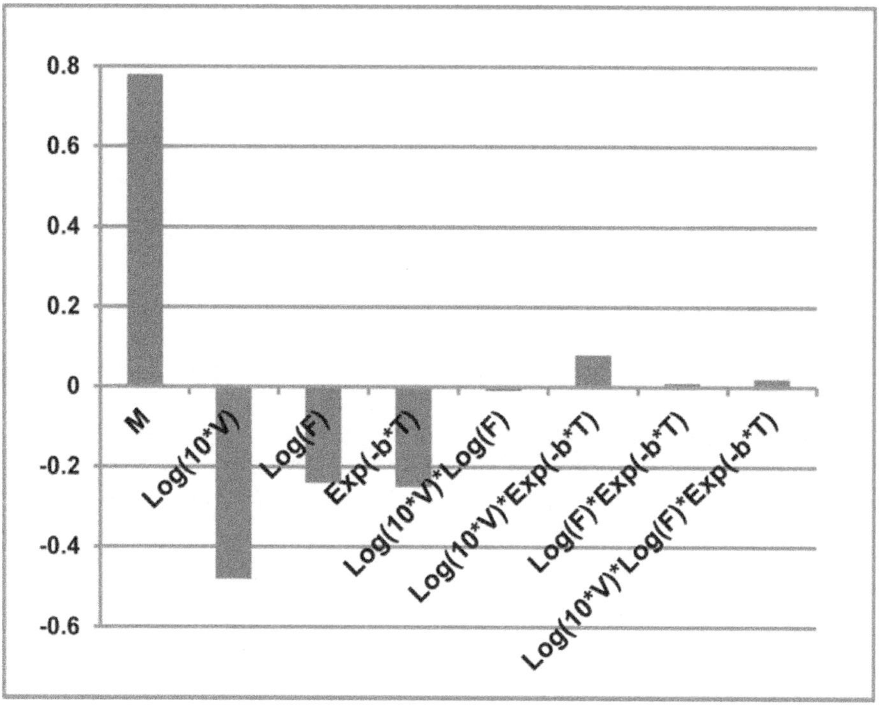

Figure 5.3. *Average effects of the stress factors and their interactions with DoE*

In this case, the model parameters are estimated from the eight DoE experiments and the 10 additional experiments needed for RSM which constitutes the learning set. The remaining 14 experiments compose the test set. The effects of the stress factors are shown in Figure 5.4, which also points out that the most influential contributions on the lifespan model are the voltage, temperature and their interaction, plus the quadratic term of the temperature. The average, maximum and minimum relative errors between the predicted and measured lifespan are provided in Table 5.6.

The RSM model shows a better performance with regard to the computed relative errors compared to the DoE model. It is important to point out that the presented methodology based on DoE and RSM methods is efficient regardless of the studied insulating material. The DoE method establishes a simple lifespan model with only eight experiments. The extended RSM model improved the model accuracy in the case of PEI films also.

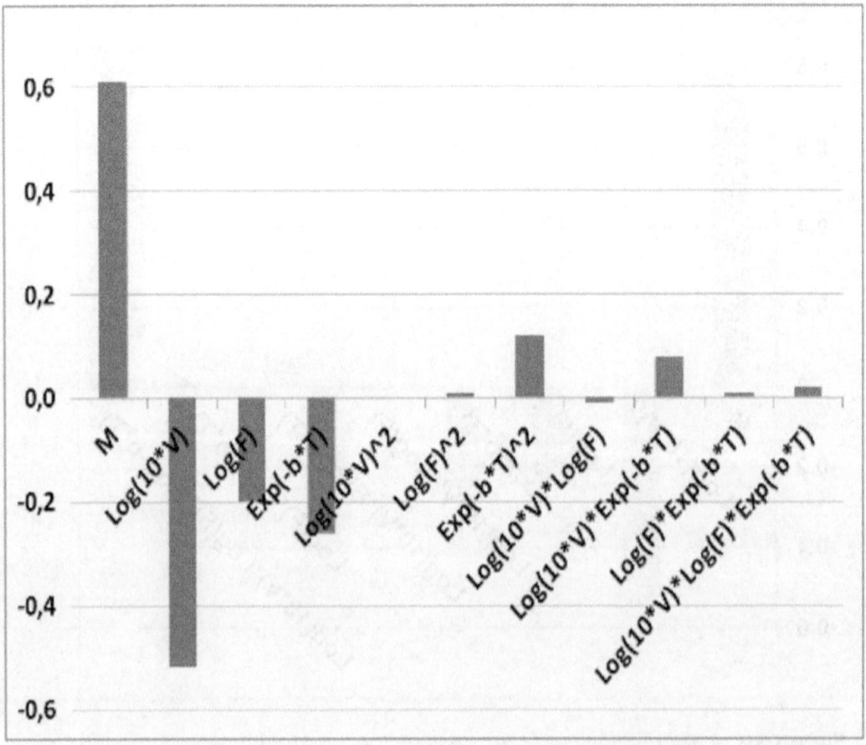

Figure 5.4. *Average effects of the stress factors and their interactions with RSM*

	Model M_1 with DoE	Model M_2 with SRM
Average error	18%	17%
Maximum error	136%	60%
Minimum error	0%	0.45%

Table 5.6. *Relative errors between the predicted and measured lifespan (model M_1)*

In the case of RSM, the lifespan model M_2 is the same as for DoE with three extra quadratic terms

$$\log(L) \approx \text{model}_{\text{DoE}} + I_{VV} \cdot \log^2(10V) + I_{FF} \cdot \log^2(F) + I_{TT} \cdot \exp(-2b \cdot T). \quad [5.8]$$

5.2.3.2. *Application to OLEDs*

Lifespan modeling of OLEDs is a complex task, as it involves different potentially interacting factors. So far this problem has not been addressed in detail, hence new parametric models based on DoE, for calculating the lifespan of OLED under thermal and electrical stresses are proposed. In the literature, luminance degradation is studied as a lifespan indicator [ZHA 12, ZHA 14, KIM 17]. The proposed models include two accelerated main factors: the temperature, following an Arrhenius law, and the current density, following an inverse power law in accordance with existing single factor models in the literature. Thus, a logarithmic transformation of lifetime, a logarithmic transformation of the current density and an inverse temperature transformation in K are applied. OLED commercial lighting sources (Philips Lumiblade OLED Panel GL55) are tested.

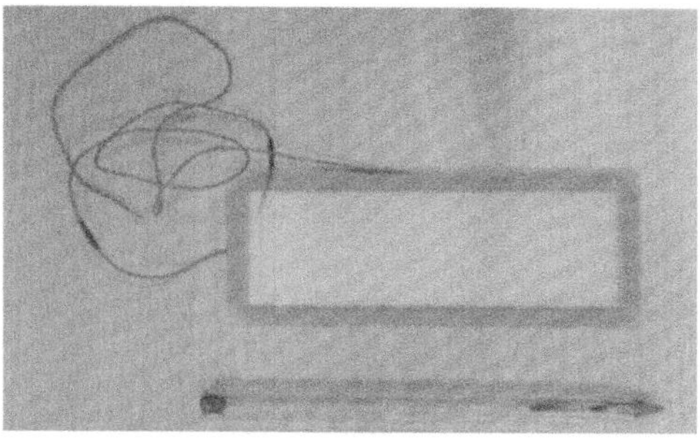

Figure 5.5. *Philips GL55 OLED*

An innovative experimental device was developed at the Laplace Laboratory to apply thermal and electrical stresses with high accuracy, homogeneity and stability, to the OLED and to simultaneously characterize *in situ* electrical and photometric characterizations. This test bench is described in [SAL 19].

Figures 5.6 and 5.7 show the evolution of the relative luminance (ratio of the measured luminance on the initial luminance) for the different aging tests with different rates of degradation of the luminance. The percentages 60, 65, 70, 75, 80 and 85% on the vertical axis measure the luminance decay with respect to the initial value measured at $t = 0$. The levels of the two main stress parameters are listed in Table 5.7. The combined electrical and thermal stress tests were designed to be able to construct several first- and second-order models with interactions. Experimental points 1–9 of Table 5.7 plotted in

Figure 5.8 form a three-level experimental plan: with two factors (temperature T and current density J) for which $3^2 = 9$ experiments are necessary.

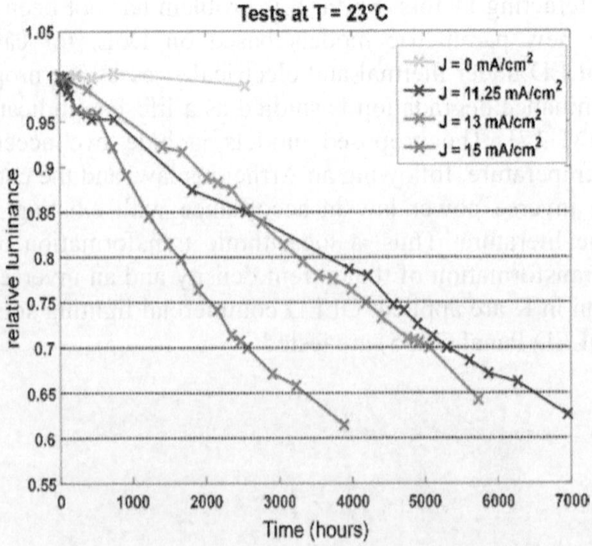

Figure 5.6. *Test results: relative luminance at T = 23°C for different current densities*

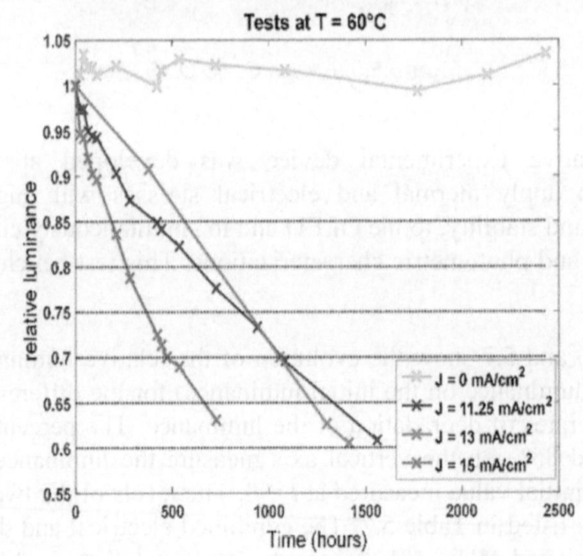

Figure 5.7. *Test results: relative luminance at T = 60°C for different current densities*

Exp. No.	Constraints		Measured lifespan (in hours)				
	J (mA/cm²)	T (°C)	L_{85}	L_{80}	L_{75}	L_{70}	L_{65}
1	11.25	23	2543	3660	4562	5298	6489
2	13	23	2566	3325	4191	5063	5644
3	15	23	1225	1654	2051	2468	3343
4	11.25	40	1234	1657	2166	2917	3488
5	13	40	860	1192	1517	1955	2702
6	15	40	949	1266	1576	1872	2377
7	11.25	60	423	628	855	1082	1331
8	13	60	567	733	893	1055	1221
9	15	60	187	270	367	471	665

Table 5.7. *OLED accelerated aging test configurations*

From this arrangement of the experimental points, we obtain

– a first-order model with interactions as in [5.1] by using the classical plan 2^2 with the extreme levels –1 and +1 (experiments 1, 3, 7 and 9), called model M_1;

– four first-order models with interactions from the four internal plans, called M21–M24;

– a second-order model, with interactions as well, based on expression [5.3]. This last model is based on all the experimental points: the extreme factorial plan 2^2 (experiments 1, 3, 7 and 9), the four axial points (experiments 2, 4, 6 and 8) and the only central point (exp. no 5). It is called model M_2.

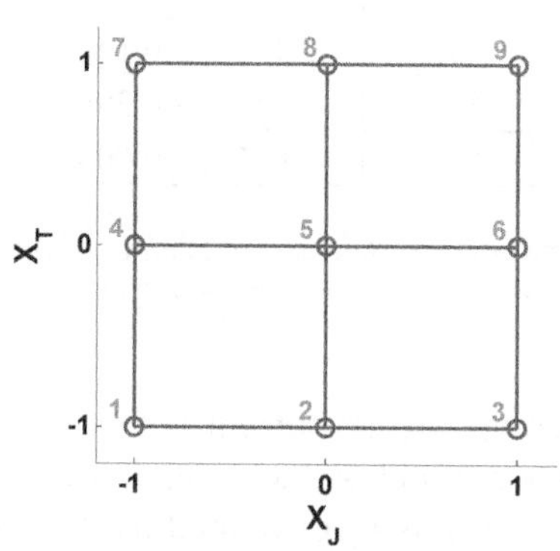

Figure 5.8. *Experimental points of Table 5.7 in 2D space*

Table 5.7 presents nine experiments where the levels of the factors are denoted by X_J and X_T for the current density and the temperature, respectively. In all the following models, L_x denotes the lifespan measured at x% of the initial luminance in hours. X_J and X_T are the respective levels of log(J) and $1/T$, as shown in Table 5.7. The first-order model with interactions is given by expression [5.9]. These experiments are the four extreme points of the square in Figure 5.8.

$$M1 = Log(L_x) = M + E_J X_J + E_T X_T + I_{JT} X_J X_T \quad [5.9]$$

The model coefficients, obtained from OLS, are given in Table 5.8 for the two studied luminance L_{85} and L_{70}, measured at 85 and 70% of the initial luminance in hours, respectively. For this first model, the test base is composed of experiments 2, 4, 5, 6 and 8 which are not part of the learning set. Tables 5.8 and 5.9 give the relative errors between the measured log(L_{70}) and log(L_{85}) and their corresponding prediction by model M_1. The factorial plan with three equidistant levels (–1, 0 and 1) can be the basis of a new DoE, including quadratic effects on each of the factors, in order to introduce some nonlinear effects. The model is given by [5.10] according to the surface response method [ESC 06]. Experiments 1–9 in Table 5.7 were used to estimate the coefficients:

$$M4 = Log(L_x) = M + E_J X_J + E_T X_T + I_{JJ} {X_J}^2 + I_{TT} {X_T}^2 + I_{JT} X_J X_T \quad [5.10]$$

The quadratic effects of current density and temperature are I_{JJ} and I_{TT}, respectively. Table 5.9 lists all the coefficients estimated by OLS.

	L_{70}	L_{85}
M	3.206	2.848
E_J	–0.173	–0.168
E_T	–0.352	–0.399
I_{JT}	–0.007	–0.009

Table 5.8. *Estimated normalized coefficients of model M1 built from extreme points of the first-order two-factor DoE*

Temperature, when compared to current density, is the most influential parameter with the strongest effect. A large quadratic effect of both factors is also evidenced, and the values of I_{JT} estimated by models M_1 and M_4 are the same. Figure 5.9 shows the comparison between the lifespan models M_1 to

M_4 for luminance degradation L_{85}. Note that most of the errors remain under 25%.

	L_{70}	L_{85}
M	3.396	3.083
E_J	–0.148	–0.131
E_T	–0.348	–0.375
I_{JJ}	–0.079	–0.122
I_{TT}	–0.084	–0.075
I_{JT}	–0.007	–0.009

Table 5.9. *Estimated coefficients for model M4*

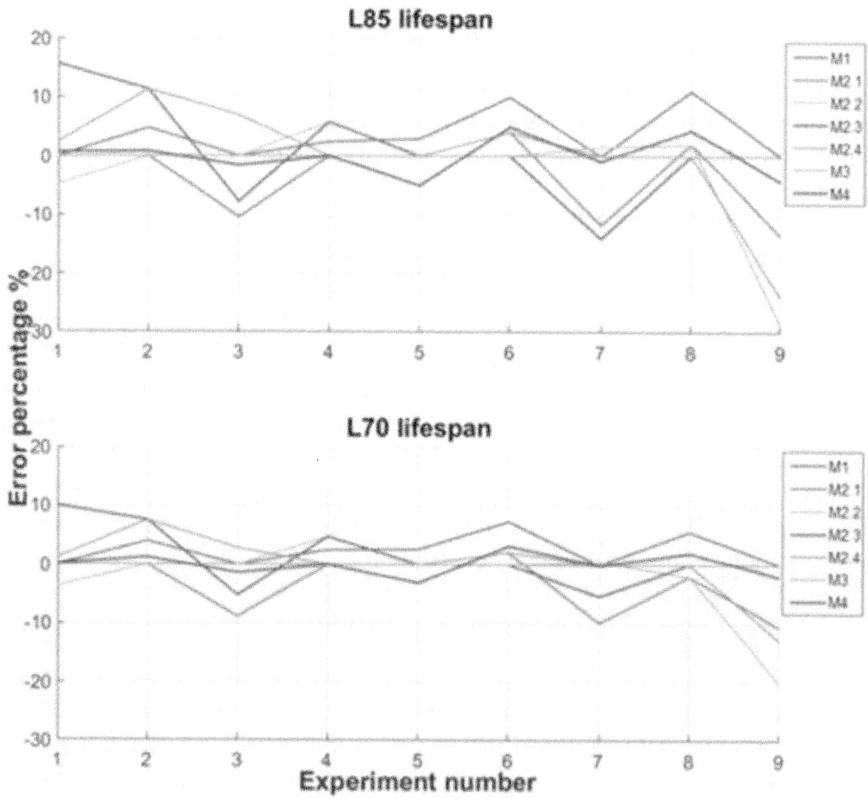

Figure 5.9. *Comparison between the lifespan models M1–M4 for luminance L85 and L70 degradation*

These models can be classified with respect to their accuracy, from the least to the most: $M_{2.x}<M_1<M_3<M_4$. The global conclusion is that nonlinear models are more accurate than linear models. $M_{2.x}$-type models, though all the four models created from specific parts of the experimental plan, are less accurate even though their maximum error remains under 30%. Another possible conclusion is that lifespans L_{85} and L_{70} do not follow exactly despite having some similarities. It can be assumed that the degradation of the OLEDs is a time-varying process, even for constant stress levels.

5.2.4. *Factor effect significance*

When computing DoE models, some of the effects may be close to zero. We may then wonder whether these effects are significant or only due to the variability of the data, that is, caused by noncontrolled factors. This question can be answered using ANOVA, provided the data can be assumed normally distributed for each factor and independent.

Let V_A be the variance due to the specific factor A and V_r the variance of the complete data set, also called the residual variance. If the factor is not significant, the variance V_A is very close to V_r. In the opposite case, the variance V_A is significantly higher than V_r. ANOVA is processed using $F_{exp} = V_A/V_r$ and tested using a Fisher–Snedecor test. This method tests the equality of variances. The null hypothesis corresponds to the nonsignificance of factor A. In this case, the ratio F_{exp} should be smaller than the threshold F_{lim}. Factor A is considered as statistically significant if F_{exp} is greater than F_{lim}. The threshold F_{lim} is defined according to the table of upper critical values of the F-distribution depending on the degrees of freedom of the data and on the test significance level (generally fixed at 5%). For each configuration, a residue r_{ij} is computed as the difference between the lifetime obtained for an experiment j and the average lifetime obtained for this particular configuration i. The residual variance V_r is computed for all configurations i and repetitions j according to

$$V_r = \frac{\Sigma r_{ij}^2}{dof_r} \qquad [5.11]$$

where dof_r is the number of degrees of freedom for the residues. If we consider a DoE with N experiments and k repetitions for each test, the degrees of freedom for each test will be equal to k–1 and so the number of degrees of freedom for the residues will be $dof_r = N(k-1)$.

For each factor A, a variance V_A is computed as

$$V_A = \frac{n \sum {E_A}^2}{dof_A} \quad [5.12]$$

where n is the number of tests when A is to a certain level, E_A is the effect of the factor A and dof_A its number of degrees of freedom. The number of tests n can be computed as the total number of tests Nk divided by the number of levels l_A for the factor A. The number of degrees of freedom is equal to the number of levels of the factor minus 1. The variance V_A can then be expressed as

$$V_A = \frac{N.k \sum {E_A}^2}{l_A(l_A-1)} \quad [5.13]$$

The square of the effect is summed as many times as the number of levels. In the case of an interaction between two factors A and B, the variance V_{AB} is computed as

$$V_{AB} = \frac{n \sum {I_{AB}}^2}{dof_{AB}} \quad [5.14]$$

with I_{AB} the effect of the interaction between A and B. The number of repetitions n is calculated as $N/l_A l_B$ and the number of degrees of freedom dof_{AB} is equal to the product of the number of degrees of freedom for A and B. The expression of V_{AB} is then

$$V_A = \frac{N.k \sum {I_{AB}}^2}{l_A l_B (l_A-1)(l_B-1)} \quad [5.15]$$

The square of the interactions is summed as many times as there are possibilities of combination between the levels of the different factors. For example, if A and B are both two-level factors, there are 4 (=2^2) possible combinations ([A+; B+], [A+; B–], [A–; B+], [A–; B–]).

5.2.5. *Multilinear regression*

This section proposes MLR [SAL 15] to model the lifespan as a function of the stress factors. The most common regression method is the linear regression. Nevertheless, as a multistress modeling problem is treated, simple linear regression cannot be used. The regression method that solves this problem is the MLR method [WEI 05]. Contrary to DoE and RSM, the

MLR operates on the real stress values instead of their levels. The MLR model is more flexible. Indeed, DoE or RSM can be seen as regression methods taking into account nonlinear relationships between the stress variables. However, these two methods impose a DoE with a fixed number of variables and of experimental data points. On the contrary, the MLR model can be either linear (constant terms and linear terms), linear with interactions (constant, linear and interaction terms), quadratic (constant, linear, interaction and squared terms) or pure quadratic (constant, linear and squared terms). Moreover, the MLR method does not include any consideration on the experiment organization: the experiments that make up the learning set for MLR can be chosen randomly which allows to focusing on some lifespan ranges. Finally, a bootstrap technic allows us to assess the model validity in the case of reduced size data sets. Although very generic, the method is illustrated in the particular case of insulation materials. As for DoE and RSM, the proposed MLR model relates the logarithm of the lifespan to the logarithm of the electrical stress factors (frequency and voltage) and to the temperature and possibly to terms defining the interactions between these factors.

Principle

The MLR models the relationship between the predicted variable or response Y, represented by n observations, and p independent predictor variables ($n > p$) as follows:

$$Y = \beta_0 + \sum_{i=1}^{p} \beta_i \cdot X_i + \varepsilon, \qquad [5.16]$$

where X_i is a ($n\times 1$) vector representing the ith column of the variable matrix X ($i = 1, \ldots, p$), β_0 and β_i ($i = 1, \ldots, p$) are the model coefficients to be estimated, ε is the ($n\times 1$) model error vector and Y is the ($n\times 1$) vector of the experimental measures. Note that the elements of matrix X are the real values of each factor instead of normalized levels as in the case of DoE and RSM methods. A more compact expression can be obtained by adding an ($n\times 1$) vector of ones as the first column of the matrix X:

$$Y = \sum_{i=0}^{p} \beta_i \cdot X_i + \varepsilon. \qquad [5.17]$$

The MLR model parameters are estimated by OLS. The sum of the squared differences between the observed and predicted values is minimized under several assumptions [FRE 09, WEI 05]:

– Full rank X: the p columns of matrix X are linearly independent,

– Nonstochastic X: errors are uncorrelated with the predictors,

– Uncorrelated error: errors are random and uncorrelated,

– Zero mean error: the expected value of the error term is zero,

– Homoscedasticity: the error variance is constant,

– Normality: the error is normally distributed. This assumption must be satisfied for conventional tests of significance of coefficients and other statistics of the regression equation to be valid.

Provided these assumptions are satisfied, the OLS estimates are optimal in the sense that they are unbiased, efficient and consistent. MLR performs the derivation of the least-square estimator of the model coefficients as in [FRE 09]:

$$\hat{\beta} = (X^t \cdot X)^{-1} \cdot X^t \cdot Y. \qquad [5.18]$$

Performance criteria

Regression model performance is evaluated through the following statistical criteria [FRE 85], [WEI 05]:

1. The standard error σ is the square root of the empirical variance of the regression residuals computed as

$$\sigma^2 = \frac{1}{n-p-1}\sum_{i=1}^{n}(\hat{y}_i - y_i)^2, \qquad [5.19]$$

where $\hat{y}_i$ is the predicted response value and y_i the corresponding measured value.

2. The coefficient of determination R^2 represents the proportion of variance explained by the regression model and is computed as

$$R^2 = \frac{\sum_{i=1}^{n}(\hat{y}_i - \bar{y})^2}{\sum_{i=1}^{n}(y_i - \bar{y})^2}, \qquad [5.20]$$

where $\overline{y}$ is the average of all measured values y_i. However, the R^2 value may increase when the predictors are added to the model. The adjusted determination coefficient is then introduced to compensate for this artificial increase, by taking into account the number of predictors p and of data points n [5.21]:

$$R^2aj = \frac{\sum_{i=1}^{n}(\hat{y}_i - \overline{y})^2 / (n-p)}{\sum_{i=1}^{n}(y_i - \overline{y})^2 / (n-1)}. \qquad [5.21]$$

3. Confidence interval for the estimated coefficients: if the regression assumptions on the residuals are satisfied, including the normality assumption, then the sampling distribution of the estimated regression coefficients is normal with variances proportional to the residual mean square error (MSE). These variances are the diagonal elements of the matrix obtained by

$$\mathrm{var}(\hat{\beta}) = \sigma^2 (X^T X)^{-1}. \qquad [5.22]$$

A (1–α)% confidence interval (CI) of a coefficient β_j can be expressed as

$$CI_{1-\alpha}(\beta_j) = \left[\hat{\beta}_j - t_{(1-\alpha/2)}\sqrt{\sigma^2 (X^T X)^{-1}}; \hat{\beta}_j + t_{(1-\alpha/2)}\sqrt{\sigma^2 (X^T X)^{-1}}\right], \qquad [5.23]$$

where $t_{(1-\alpha/2)}$ is the (1–α/2) percentile of Student(n–p–1) distribution.

4. QQ-plots allow us to check the normality of the regression residues. A QQ-plot (Q stands for quantile) compares two probability distributions by plotting their quantiles against each other. In the context of regression models, normal distribution is to be verified, thus, residue quantiles are compared to the corresponding theoretical normal quantiles. If the two distributions are identical, the QQ-plot follows the 45° line $y = x$.

Application to insulation materials

The performance of MLR is presented and compared to the ones of DoE and RSM in the case of insulation materials. Main factors, interactions and quadratic effects are included in the regression model defined by

$$Y \approx \sum_{i=0}^{10} \hat{\beta}_i \cdot X_i. \qquad [5.24]$$

The $\hat{\beta}_i$'s are the model parameters and the X_i's are the real values of the so-called explanatory variables 1, log(10*V*), log(*F*), exp(–*bT*), log^2(10*V*), and so on in the same order than in the RSM model. Since 11 parameters are considered, at least 11 experiments are required to estimate our regression model.

In the first step, these experiments are randomly chosen among the available 32 experiments to assess their performance on the global lifespan range. The learning set is composed of experiments1, 3, 7, 8, 9, 12, 23, 24, 26, 29 and 30. The average, maximum and minimum relative errors between the predicted and measured lifespans are shown in Table 5.10. With this model constructed with 11 randomly selected experiments, we obtain lower errors than those obtained with DoE or RSM.

Average error	35%
Maximum error	152%
Minimum error	0%

Table 5.10. *Relative errors between the predicted and measured lifespans (model M3)*

Note, however, that, as for DoE and RSM, the maximum error of 152% for MLR occurs for a short lifespan of 2.22 minutes. The errors obtained with DoE, RSM and MLR with respect to measured lifespan reveals that the highest errors are obtained for very short lifespan, as shown in Figure 5.10. We can assume that very short lifespan results from a particular degradation regime and thus can be seen as outliers in the need for long lifespan models. Therefore, a given model cannot fit long and short lifespans simultaneously. Except short lifespan, the models outperform in the long lifespan domains which is obviously the most interesting one.

To build a model with acceptable performance for the lifespan range of interest (e.g., long lifespan), the experiments corresponding to very short lifespan must be removed from the learning set. However, DoE and RSM methods require a fixed design of the learning set that may contain long and short lifespan. From this point of view, MLR shows more flexibility. Some ranges of interest from our data set can be chosen to generate the MLR models, particularly, in the long lifespan range. It can be seen from Figure 5.10 that model errors significantly decrease for log(*L*) > 0.5, which corresponds to a lifespan $L \cong 3$ minutes. As a first attempt, this value is used in the following as a threshold to define short and long lifespan intervals. Regression is performed for different lifespan ranges, using all experimental data in the considered range as a learning set. In particular, the six repetitions are used in the Y vector instead of using only the median value per experiment. This procedure aims at enriching the learning set. Consequently, the X matrix has to be reshaped. Since the six measured lifespan per experiment are subjected to the same constraints (V, F and T), each line of the original X matrix must be repeated identically six times. Finally, the obtained models are compared based on the criteria defined in section 4.2.4.2.

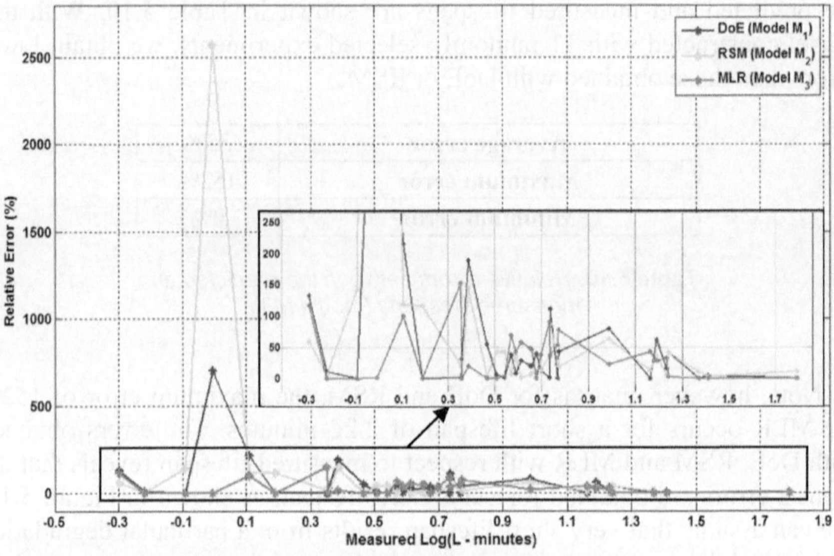

Figure 5.10. *Model (DoE, RSM and MLR) relative error distribution according to the measured lifespan*

Model (a) with all experiment data points

First, all data points (32×6 = 192) are used to construct an MLR model (Model (a)) regardless of the lifespan range. Figure 5.11 displays the normal QQ-plot of the regression residuals. As only 5% of the total number of observations does not fall on the straight line (note that, as expected, they correspond to small extreme values, see the left-hand side of Figure 5.11), residue normality can be assumed as globally verified. Thus, the upper and lower confidence interval bounds of the model coefficients can be calculated (Figure 5.12). This figure points out that most of these coefficients have large confidence intervals, which means that their variances are too high and therefore they cannot be used to represent a unique model over the entire lifespan domain.

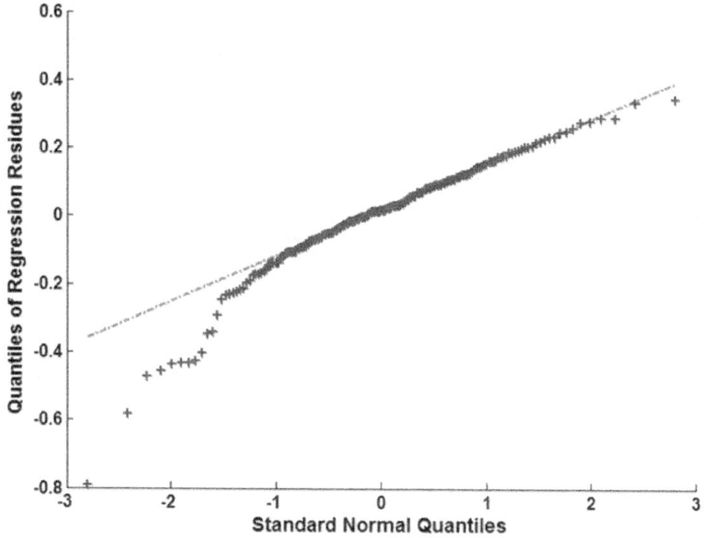

Figure 5.11. *QQ-plot of model (a) residues*

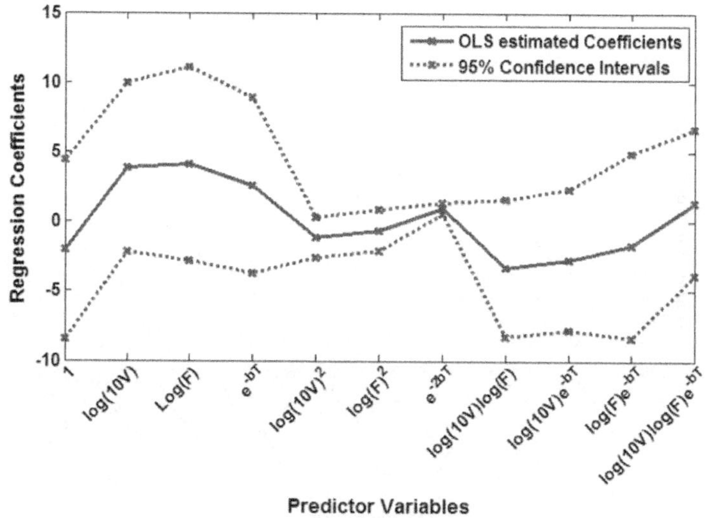

Figure 5.12. *Model (a) regression coefficients with 95% CI*

Indeed, the average relative error between the estimated and measured values is 47%. However, the average relative error for high lifespans is 14% only, meaning that the highest errors are mostly located in the short lifespan range.

Model (b) only with long lifespan data points (L > 3 minutes)

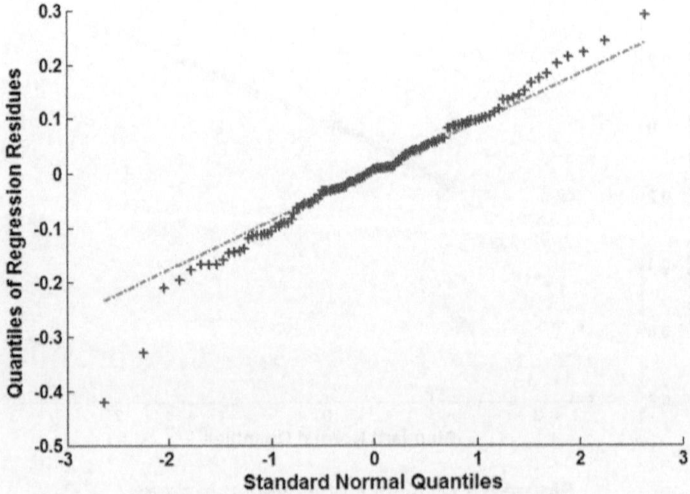

Figure 5.13. *QQ-plot of model (b) residues*

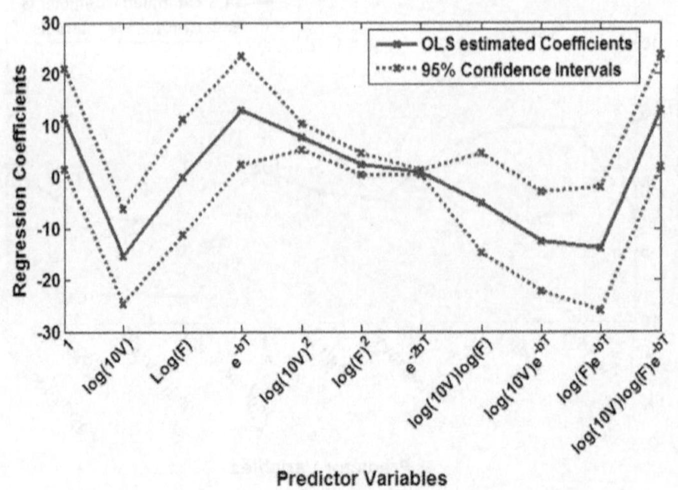

Figure 5.14. *Model (b) regression coefficients with 95% CI*

Second, only long lifespan ($L > 3$ minutes) are taken into account to generate the regression model, and hence the number of data points is therefore reduced to 121 instead of 192. Figure 5.13 shows a better fitting to normal distribution, in comparison with Figure 5.11. Figure 5.14 also shows smaller CIs for the model coefficients. This guarantees a better confidence level for the global model in this lifespan range ($L > 3$ minutes) with respect to Model (a).

Finally, the average relative error between the estimated and measured values in this model is 10%, which is significantly reduced with respect to the average error in Model (a) (47%).

Comparison of model (a) and model (b) with radar plots

A radar plot can be used to graphically illustrate the comparison between these two models. A radar plot allows us to observing m relevant performance criteria at a glance. It is composed of m axes having equal lengths, starting from the same point and graduated according to the value of each criterion. A line is drawn connecting the values on each axis, thus describing the global performance. In the case of the study, eight criteria are considered: the standard error σ, the adjusted determination coefficient R^2, the maximum relative error, the maximum relative error between the estimated values and the mean value of the six repeated measurements on each experiment, the mean relative error, the mean relative error between the estimated values and the mean value of the six repeated measurements, the proportion of data points having a relative error higher than 20% and finally the same when the error is computed between the estimated values and the mean value of the six repeated measurements. Note that criteria are considered in mean for a given experiment in order to analyze the model performance in average, regardless of the dispersion of the measurements.

For a more comprehensive representation, $1-R^2$ is considered instead of R^2 so that good performance amounts to a small value of the criterion as it is the case for the other ones. The radar plots of the two models are shown in Figure 5.15.

This analysis shows a significantly better performance when only the long lifespans are accounted for in the regression model. It confirms, once again, our hypothesis that very short and long lifespans cannot be modeled simultaneously. In conclusion, a better performance is achieved by considering only long lifespans. Fortunately, it is also our domain of interest since our objective is to validate a lifespan model that can be applied to nominal constraints leading to long (or even very long) lifespans.

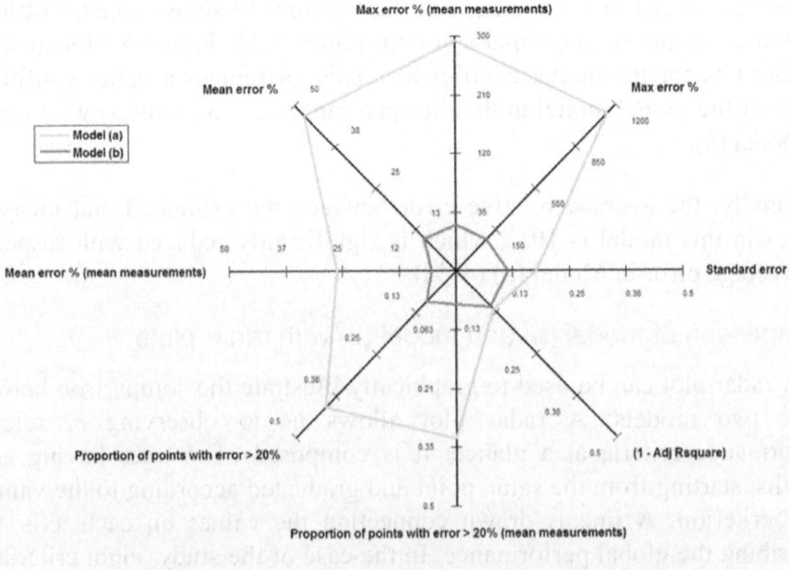

Figure 5.15. *Radar plot of models (a) all lifespans and (b) lifespans L > 3 minutes versus eight quality criteria*

Bootstrap method

As previously demonstrated, graphical tools such as QQ-plots and goodness-to-fit tests can be used to validate the underlying hypothesis required for OLS coefficient estimation. However, with very small training sets, it might be difficult to assess some hypotheses regarding statistical properties. In this case, the performance of coefficients can be estimated using nonparametric bootstrap method. The bootstrap method aims at inferring statistical properties of an estimator from a nonparametric resampling of the data [EFR 93]. In the case of regression, a high number (B, $50 < B < 200$) of replications of model coefficients β^* must be generated. Two resampling methods exist for regression models [EFR 93]:

– *Bootstrap on xy pairs:* B models are estimated by making random sorts with retrieval of pairs (X^*,Y^*) in the original (X,Y) sample.

– *Bootstrap on residuals:* B models are estimated by creating new response values Y^* from random sorts with retrievals in the original residual sample.

Using the obtained bootstrap replications β^*, their standard errors SE(β), their confidence intervals CI(β) and their statistical significance (Student's test p-values) can be computed as shown in Figures 5.16 and 5.17. As it does not require any assumption on the underlying distribution, the bootstrap method is interesting for making statistical inference on small sizes. The bootstrap method is applied on DoE coefficients with $B = 200$ and using the two resampling methods. Figures 5.18 and 5.19 show, respectively, the CI of the coefficients and p-values of Student's test obtained with the bootstrap method, as well as those computed under residual normality assumption.

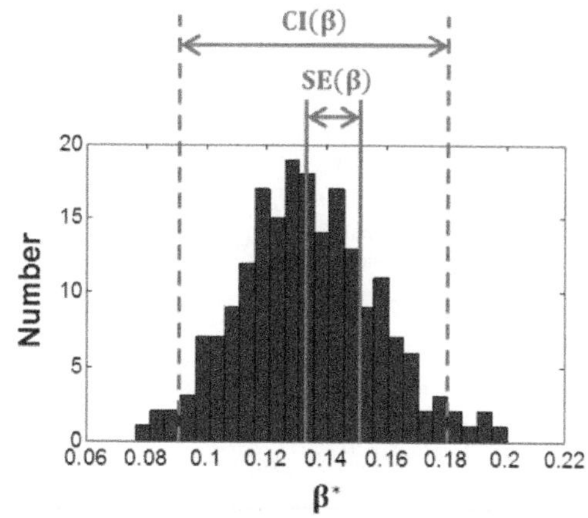

Figure 5.16. *Bootstrap SE and SI*

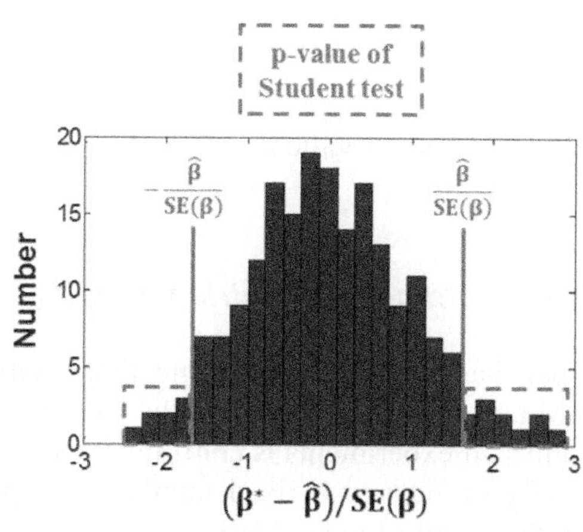

Figure 5.17. *Bootstrap p-values of Student's test*

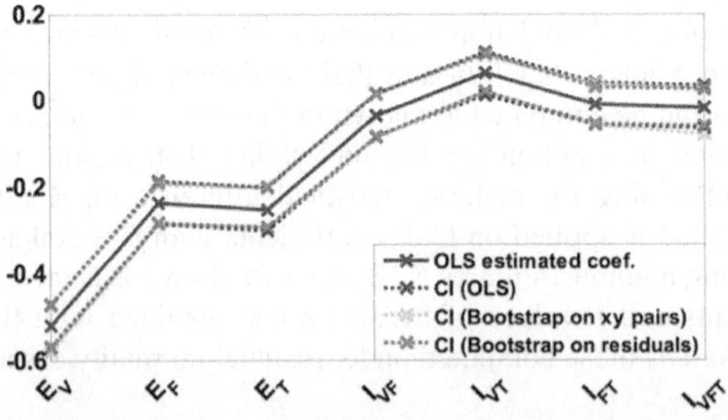

Figure 5.18. *CIs for DoE model coefficients under residual normality assumption and with bootstrap*

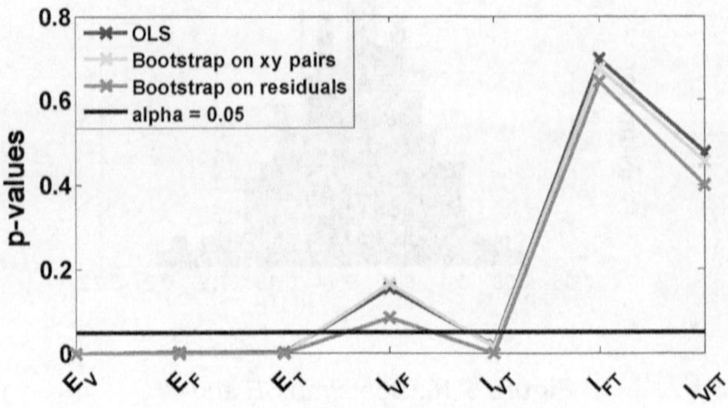

Figure 5.19. *P-values of Student's test applied to DoE model coefficients under residual normality assumption and with the bootstrap method*

Finally, the bootstrap method validates the underlying hypotheses made for parameter estimation.

5.2.6. *Conclusion regarding parametric methods*

Insulation lifespan has been modeled using three statistical parametric methods: DoE, RSM and MLR. The ability of the DoE and RSM to organize and to limit the number of experiments is confirmed. However, these models impose a fixed configuration and a fixed number of experiments in their learning sets. From this point of view, MLR method shows more flexibility. After a first analysis of DoE, RSM and MLR models, the studied relative errors reveal that these models have a better predictability in the long lifespan range. Moreover, a detailed regression analysis was performed to confirm that short and long lifespans cannot be modeled simultaneously with a unique model involving the considered number of parameters (11 parameters). Hence, regression methods can help analyze different regions inside the studied domain and to eliminate outliers that might be due to a different aging mechanism.

5.3. Nonparametric models

Parametric models presented in section 5.2 allow for the quantification of aging factors effects and their interactions on lifespan, according to a multilinear relation that is predefined. The number of variables for a complete parametric model depends on the number of aging factors: for k factors, the number of variables, including the interactions and quadratic terms, is $(k+1)(k+2)/2$. So, when the number of factors increases, the lifespan models will be more complex and difficult to interpret. Moreover, the interactions are introduced as independent explanatory variables through the product of the corresponding factors. This choice has no physical justification. Therefore, the interpretation of the resulting coefficients is not evident.

In this section, lifespan modeling is considered a nonparametric problem in order to obtain models whose form is not predefined, but determined directly from the data. This way, it is possible to address the drawbacks of parametric models: to better understand the interactions between the factors in the first-order models and to increase the prediction quality of the second-order models. Nonparametric regression trees (RTs) present the first alternative approach to linear regression models and are especially appropriate when interactions exist between factors. A description and their application to lifespan modeling of insulating materials are given in this section. Then, a modeling based on both parametric and nonparametric models is presented. This method, called the *hybrid model*, takes advantage of the two approaches to improve the lifespan prediction.

5.3.1. *Regression trees*

Description

Classification and RTs are nonparametric predictive models based on supervised learning. Breiman introduced them in 1984 [BRE 84]. They explain the relationship between a single response variable (output) and a set of predictor variables (inputs). Outputs and inputs can be either discrete or continuous. Although RTs have been widely used for the modeling of medical data, environmental data or geological data [SUT 11, DEA 00, FEL 13]), they have been rarely used in electrical engineering [SAL 17].

The goal of RTs is to model the relationship between a variable Y (output) and several explanatory variables (inputs). To do so, RTs recursively split the training data set into smaller and more homogeneous groups, called *nodes*. The splitting explanatory variable and the corresponding threshold value are chosen at each node to maximize the homogeneity of each resulting group. Each final group, called *leaf*, is characterized by the mean value of this specific group and a succession of conditions on the explanatory variables. The relation between the output and the inputs can then be defined as

$$\hat{Y} = \sum_{i=1}^{r} \beta_i I_i \qquad [5.25]$$

where $\hat{Y}$ is the predicted output by the RT, r is the number of leaves, β_i the mean value associated with the leaf i and I_i is a Boolean variable that is equal to 1 if the inputs are satisfying the conditions of the leaf i, and 0 otherwise.

RTs are thus piecewise constant models. For a new observation, the response can be easily predicted by following the appropriate path throughout the tree. The order of appearance of the variables in the tree allows comparing their relative importance. The algorithm and criteria for the building of RTs applied to continue data can be found in [BRE 84]. The building of RTs is generally composed of two phases. In the first phase, the algorithm recursively splits the training data set 2×2, by optimizing an error criterion. In the case of big training set, this first phase may result in an overfitting of the data, and unstable models depending too much on the training data. In this case, a second phase of pruning the model obtained in the first phase. Pruning reduces the size of RTs by removing parts of the tree that do not provide power to classify instances.

RTs can be computed by the Matlab function *classregtree.m*, which is an implementation of the Breiman's CART algorithm. This function is used here for nonparametric lifespan modeling depending on different stress factors. Unlike the parametric modeling presented in section 5.2, only the main factors are considered as inputs of the algorithm. At first, RTs can be used as a classification tool on a training data set in order to identify the most influent factors and the hierarchy between them. Second, the obtained model can be used to predict the lifespan on a test data set. The purpose is the same than with parametric model: to obtain the most relevant model with a reduced number of samples.

Example: lifespan modeling of insulated electrical twisted pairs

Here, the methodology of RTs is applied to the lifespan modeling of a 220°C thermal class insulating material that is widely used for the wiring insulation of electrical machines. Test samples are twisted pairs of 0.5-mm diameter copper wires covered of PAI, already presented in section 5.2.3.1. Owing to the low number of samples, complete RTs are built and no pruning is applied. The prediction quality is evaluated as the relative error between the lifespan predicted by the RT Y_p and the measured lifespan Y_m. The input parameters are the levels of the three main factors V, F and T (the voltage, the frequency and the temperature, respectively).

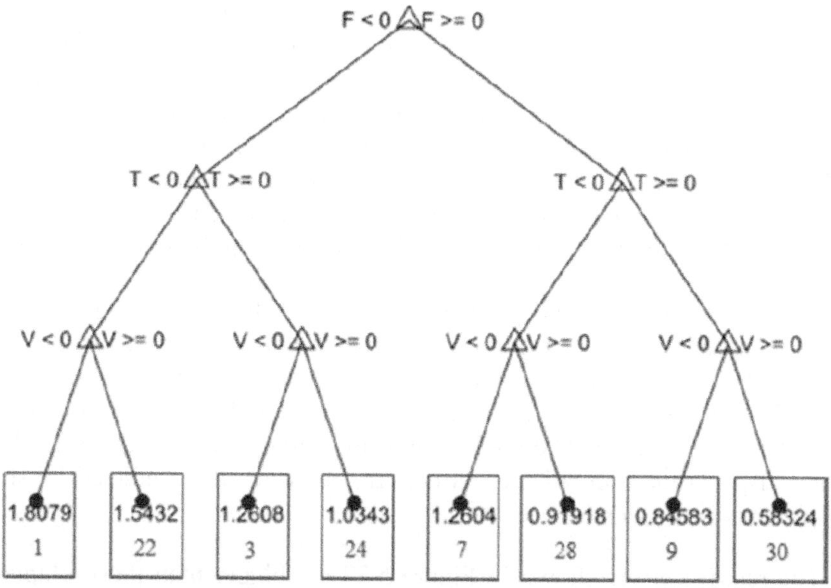

Figure 5.20. *RT 1 built from experiments 1, 3, 7, 9, 22, 24, 28 and 30*

We focus here on the test samples corresponding to the DoE training set. The first RT (tree 1) is built from experiments 1, 3, 7, 9, 22, 24, 28 and 30 of Table 5.5. In these experiments, each factor is at level −1 or +1. The results are given in Figure 5.20 where every leaf is associated with the average value of the lifespan logarithm and the corresponding experiment number. It can be noticed from Figure 5.20 that the frequency is the main splitting factor. As a comparison, this factor has the most important effect in the case of a DoE modeling on the same training set. The hierarchy between the factors is the following: frequency, temperatures and then voltage. The same order is also observed for the factor effects in the case of DoE modeling. The order between V and T is the same for each frequency zone which proves the homogeneity of the model on the range [–1; +1].

A second RT (tree 2) is built with all the 30 experiments presented in Table 5.5. The result is presented in Figure 5.21. The experiments used in Figure 5.20 are displayed in blue boxes, whereas the complementary experiments are displayed in red.

The comparison between tree 2 and tree 1 shows that the frequency is still the main splitting factor and so the most important factor. Moreover, the position of the eight experiments used for tree 1 stay the same (relatively to each other) despite the addition of the 22 new experiments. The addition of supplementary experiments modifies the order of the factors in the different nodes compared to tree 1. The left subtree remains homogeneous ($F<0.5$), whereas the behavior of the right part ($F>0.5$) is different. The left subtree is indeed subdivided into three temperature zones and the same factor order is observed in each zone (F then V). This is different in the right subtree which means that there are two different models depending on the frequency zone.

In order to evaluate the prediction quality of lifespan modeling using RTs, the model obtained in Figure 5.21 (tree 2) is tested on a test data set composed of random experiments labeled as experiments 41–64 in Table 5.5. Only tree 2 is tested here because the model is more complete due to the large training set. The prediction results are given in Figure 5.22. This figure displays the normalized relative error between the predicted lifespan and the experimental measure.

The relative errors shown in Figure 5.22 are bigger than those obtained for a parametric model built from the same training set. The relative errors remain indeed smaller than 10% in the case of a parametric model (section 5.2.3), whereas the ones for the RT reach more than 20%. Figure 5.23 compares the lifespan prediction of tree 2 (green circles) with lifespan prediction obtained by a linear parametric model built on the same training set (experiments 1–30) and the main factors (V, F and T) only.

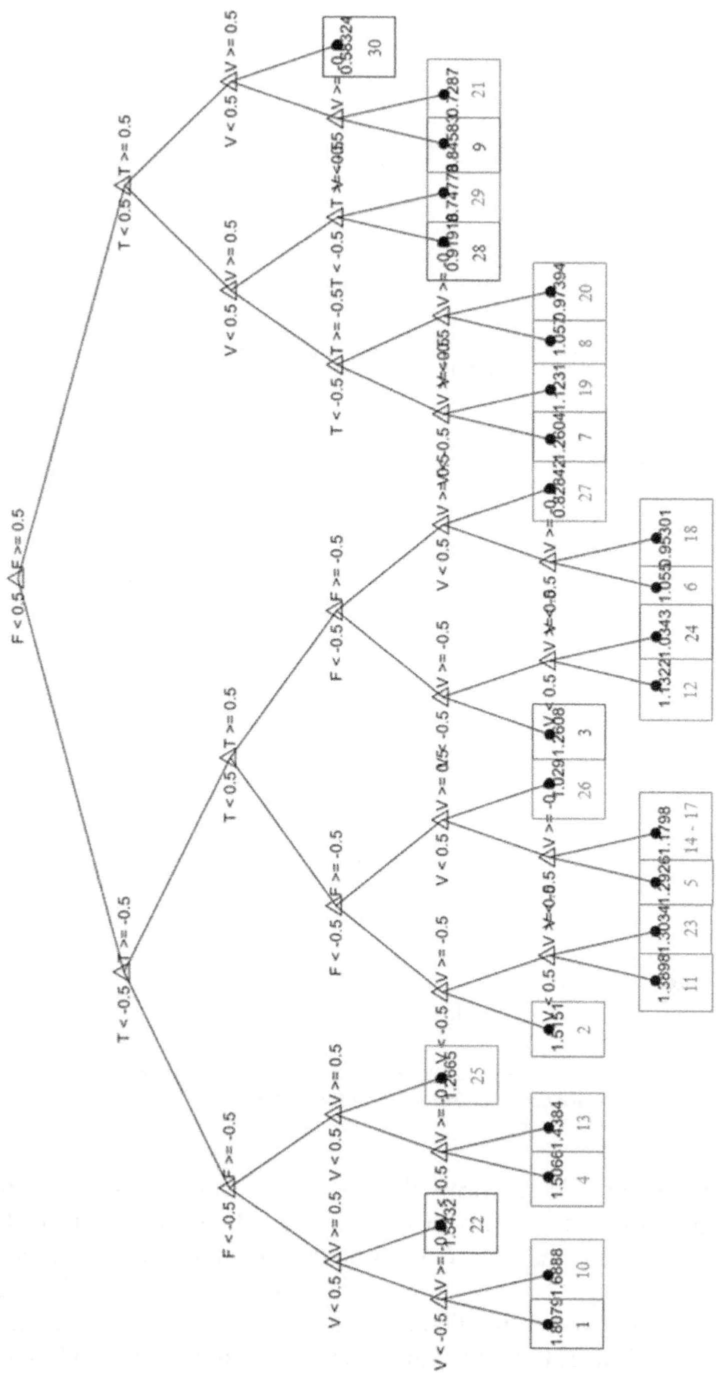

Figure 5.21. *RT 2 built from experiments 1–30*

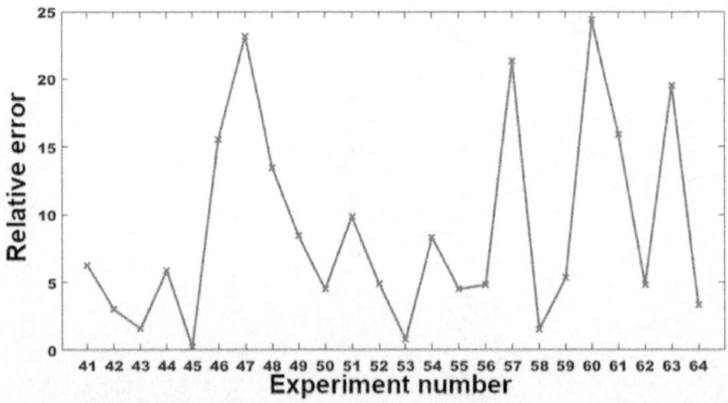

Figure 5.22. *Lifespan prediction error of tree 2*

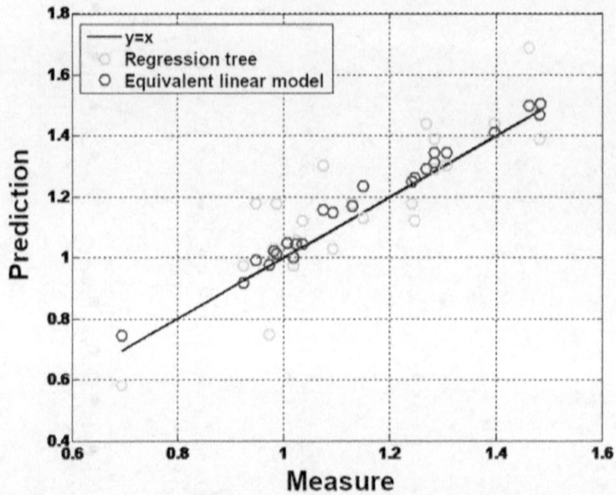

Figure 5.23. *Comparison between lifespan predictions of tree 2 and those of an equivalent linear parametric model*

Analysis of Figure 5.23 confirms that the equivalent parametric model attains better performance in terms of lifespan prediction. This result is due to the fact that RTs are piecewise constant models and thus can only predict already known lifespans, that is, lifespans that are in the training data set. RTs have intrinsically lower prediction accuracy than parametric models. Nonetheless, RTs allow us to build simple and graphical lifespan models and remain an interesting tool in order to analyze the relative importance of the main factors that can be deduced from their hierarchical structure.

5.3.2. *Hybrid models*

Motivations and application to lifespan modeling

The previous section has shown that RTs, based on a binary segmentation of the constraint domain, are a good tool to assess the hierarchical relationship between the main stress factors and the lifespan. Nevertheless, RT-based lifespan models have also proved to be limited in terms of prediction due to the fact that they are piecewise constant models. On the other hand, it has been demonstrated in section 5.2 that analytical relations between lifespan and the stress factors, their interactions and their quadratic effects can be assessed through parametric modeling. Both parametric and nonparametric models have shown advantages and disadvantages.

The effect of each factor and their interaction can be evaluated through the coefficient values related to each explanatory variable. The relative importance of each variable of the model can thus be estimated and the model can be simplified by neglecting the variables that are statistically nonsignificant. In the case of small data set, it seems that first-order models with interactions give the best results in terms of lifespan prediction. However, the number of parameters increases exponentially with the number of stress factors if we want to model their interactions and potentially their quadratic effects. So, parametric models become more and more complex with the number of variables. It also increases the size of the training data set. Moreover, the effects of interactions are often difficult to interpret.

From this perspective, nonparametric models such as RTs are more appropriated because the relationship between different variables is explicitly expressed through the hierarchical structure of the RT. The relative importance of each factor can then be deduced, along with specific zones where the factors have the same behavior (identical node structures). The structure of RTs is well suited in order to interpret the relationship between the lifespan and the stress factors, but errors are quite important when it comes to lifespan prediction, which is a major drawback. RTs are indeed piecewise constant models, and better lifespan predictions mean adding new nodes and leaves, and so a bigger training data set. So, with a similar training set, a linear model built only from the main factors reaches better prediction performance than RTs.

Hybrid models are thus proposed here in order to benefit from both modeling approaches. The purpose is to take advantage of the explicit nature of parametric model analytic expressions, along with their quality in terms of lifespan prediction, on the one hand, and to use the ability of RTs to classify experiments in different zones depending on the applied constraints, on the other hand. The proposed approach combines both parametric and nonparametric approaches in order to get a piecewise linear model, called a hybrid model (HM).

There are some models in the literature that combine parametric and nonparametric approaches, such as generalized RTs [CHA 95], the regression trunk approach [DUS 04] or the stepwise model tree induction [MAL 04]. The principle of these methods is to use RTs globally to split between the different variables and to associate a linear model to each terminal node. The proposed HM is taken from this general principle with a faster and simpler algorithmic approach that aims to obtain a model well suited for lifespan modeling.

The principle here is first to identify the most important factor and its splitting values, thanks to the RT. Then, a linear model is computed separately on the different identified zone. This is done concretely by means of dummy variables: one coefficient for each of the other factors is defined in each range of the main factor. This specific structure enables refinement of the parametric model by analyzing the life–stress relationship in each identified range. Moreover, it helps understand the existing interactions with the main factors by examining the effect of the main factor range on the coefficients of the other factor. By doing so, the prediction quality of RTs is improved.

Example: lifespan modeling of insulated electrical twisted pairs

For comparison purpose, HMs are applied here on the same data than those used in section 5.3.1 for the evaluation of RTs. Experiments 1–30 are used as a training set and experiments 41–64 as a testing set. According to Figure 5.21, frequency is identified as the most important stress factor dividing training data set into two ranges:

– A high-frequency zone (HF): $X_F > 0.5$ ($F > 11.7$ kHz);

– A low-frequency zone (LF): $X_F < 0.5$ ($F < 11.7$ kHz).

The HM can thus be written as

$$\mathrm{Log}(L_{\mathrm{m}}) = M + E_F X_F + E_{V/HF}\delta_{HF} X_V + E_{V/LF}\delta_{LF} X_V + E_{T/HF}\delta_{HF} X_T + E_{T/LF}\delta_{LF} X_T. \quad [5.26]$$

In this equation, δ_{HF} (resp. δ_{LF}) is a dummy variable equal to 1 when X_F belongs to the HF (resp. the LF) zone and 0 elsewhere. HM coefficients estimated from the set are given in Figure 5.24. In this figure, the effect of the frequency E_F is plotted in red, while those of the voltage $E_{V/\mathrm{HF}}$ and $E_{V/\mathrm{LF}}$ are in blue, and those of the temperature $E_{T/\mathrm{HF}}$ and $E_{T/\mathrm{LF}}$ in green.

It is obvious from Figure 5.24 that different coefficient values for X_V and X_T depend on the frequency zone, which confirms the interaction between the frequency and the voltage and temperature. For both zones LF and HF, the XV effect is lower than the XT one, which is coherent with the relative position in the tree 2 presented in Figure 5.21 and built on the same learning set. The HM is then tested on experiments 41–64. Figure 5.25 compares the predicted lifespan obtained with the HM to the actual experimental lifespans on the learning (blue) and testing (red) data set.

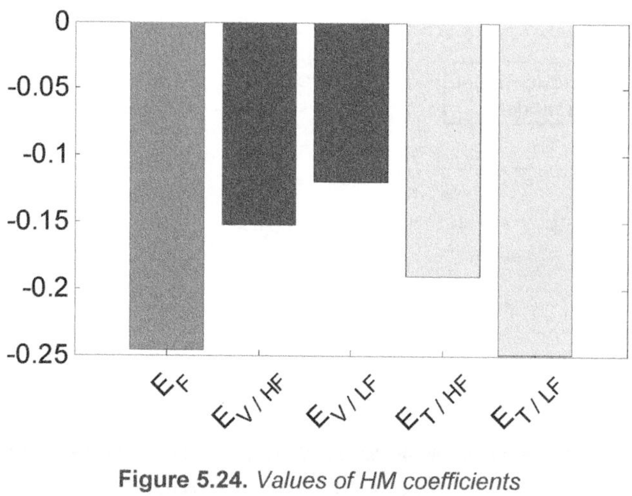

Figure 5.24. *Values of HM coefficients*

Figure 5.25 shows that the predicted lifespans are very close to the measured ones irrespective of the data set. These good results are confirmed in Figure 5.26 where the relative errors are plotted in green for each experiment. For comparison purpose, the relative errors obtained for a parametric model computed from the same learning set is plotted in red. The parametric model is a three-level full factorial design.

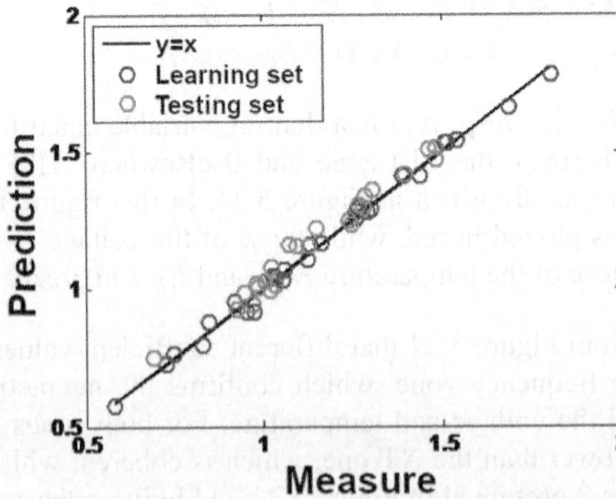

Figure 5.25. *Comparison between the predicted and measured lifespans on the learning (blue) and testing (red) data set*

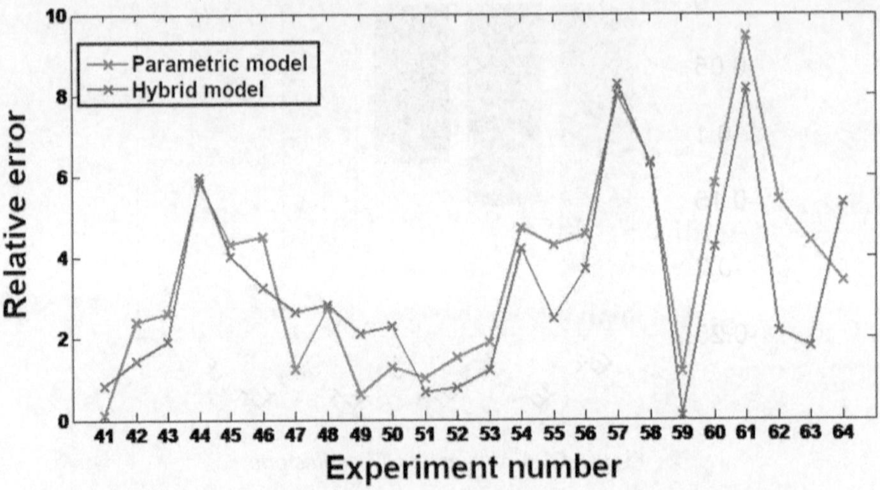

Figure 5.26. *Relative errors of the HM (green) and of a parametric model computed from the same learning set (red)*

It is clear from Figure 5.26 that the HM gives better lifespan prediction for most of the testing set with an average error of 3.4%. Moreover, HM involves a small number of variables compared with parametric models. All these variables are important and more easily interpretable than those of the parametric model. So, HM configured after identifying the different regions of the main stress factor with an RT are an interesting alternative to parametric models, especially in the case of large number of factors.

5.4. Conclusion

In this chapter, a general methodology for lifespan modeling in electrical engineering has been presented. Different models are based on statistics from accelerated aging tests. The main purpose is to provide relevant lifespan modeling at minimum experimental cost. Model accuracy is directly linked to the number of considered parameters. Electrical materials are indeed subject to several operational and environmental constraints that contribute to its degradation. Moreover, the different stress factors may interact with one another and it is not always possible to study them separately. A proper methodology in order to optimize a high number of stress factors and a low experimental cost is then needed, particularly in electrical engineering where experimental data are often limited to a few number of experiments.

First, parametric models have been presented. These models express a linear relationship between the logarithm of the lifespan and the stress factors and their interactions. Their general form is similar to a multilinear regression and they can be estimated using OLS. In a context of a reduced number of experiments, DoE and RSM methods are used in order to organize the experiments needed for the model parameter estimation. These experiments allow for the evaluation of the stress factor effects and of their interactions. The model complexity increases with increasing number of factors and the order of the model, and more experiments are then needed. The statistical analysis of parametric models allows us to assess the significance of the different terms and to neglect those who are not relevant. From this point of view, MLR shows more flexibility and also introduces quadratic terms in the model, as with RSM. Moreover, these methods can help to analyze different regions inside the studied domain and to eliminate outliers that might be due to a different aging mechanism.

The second part of this chapter focuses on nonparametric models. RTs proved to be efficient for the classification of the different experiments into different zones where the variable's behavior is the same in regards to the lifespan, and where the aging mechanisms are probably the same. Thanks to the hierarchical structure of RTs, a new interpretation of the relation between the lifespan and the stress factors is possible. However, RTs are piecewise constant and have lower prediction performance than parametric models. HMs are piecewise linear models that combine advantages of both parametric models and RTs. This approach allows for a better evaluation of the principal stress factor compared to other factors by identifying different zones for this factor that are significantly different. This way, it increases the prediction ability of RTs having a piecewise linear model rather than piecewise constant one. This approach is especially interesting in the case of numerous stress factors.

In the future, these different approaches should be studied in order to model the degradation itself rather than the lifespan which should lead to real-time prognosis of the remaining useful life of the electrical materials.

5.5. References

[ANS 07] American National Standards Institute, ANSI/NEMA MW 1000-2003, Revision 3, 2007.

[BRA 07] Bradley N., The response surface methodology, Master thesis, Indiana University, 2007.

[BRE 84] Breiman L., Friedman J., Olshen R., Stone C., *Classification and regression trees*, Wadsworth, California, 1984.

[CHA 95] Chaudhuri P., Lo W., Loh W., Yang C., "Generalized regression trees", *Statistica Sinica*, vol. 5, no. 2, pp. 641–666, 1995.

[CHR 11] Christou I., Nelms A., Cotton I., Husband M., "Choice of optimal voltage for more electric aircraft wiring systems", *IET Electrical Systems in Transportation*, vol. 1, no. 1, pp. 24–30, 2011.

[DEA 00] De'ath G., Fabricius K., "Classification and regression trees: A powerful yet simple technique for ecological data analysis", *Ecology*, vol. 81, no. 11, pp. 3178–3192, 2000.

[DUS 04] Dusseldorp E., Meulman J., "The regression trunk approach to discover treatment covariate interaction", *Psychometrika*, vol. 69, no. 3, pp. 355–374, 2004.

[EFR 93] Efron B., Tibshirani R.J., *An introduction to the bootstrap*, Chapman & Hall, New York, 1993.

[ESC 06] Escobar L., Meeker W., "A review of accelerated test models", *Statistical Science*, vol. 21, no. 4, pp. 552–577, 2006.

[FAN 13] Fang L., Cotton I., Wang Z.J., Freer R., "Insulation performance evaluation of high temperature wire candidates for aerospace electrical machine winding", in *Proc. IEEE Electrical Insulation Conf.*, pp. 253–256, 2013.

[FEL 13] Felcisimo A., Cuartero A., Remondo J., Quiros E., "Mapping landslide susceptibility with logistic regression, multiple adaptive regression splines, classification and regression trees, and maximum entropy methods: A comparative study", *Landslides*, vol. 10, no. 2, pp. 175–189, 2013.

[FIS 35] Fisher R.A., *The design of experiments*, Oliver and Boyd, Edinburgh, UK, 1935.

[FRE 85] Freeman T.G., "Selecting the best model to fit data", *Mathematics and Computers in Simulation*, vol. 27, nos 2–3, pp. 137–140, Apr. 1985.

[FRE 09] Freedman D.A., *Statistical models: Theory and practice*, revised edition, Cambridge University Press, 2009.

[GUA 11] Guastavino F., Ratto A., Torello E., Biondi G., Loggi G., Ceci A., "Electrical aging tests on different nanostructured enamels subjected to severe voltage waveforms", in *Proc. IEEE SDEMPED*, pp. 278–282, 2011.

[ILL 14] Illias H.A., Tunio M.A., Abu Bakar A.H., Mokhlis H., Chen G., "Partial discharge behaviours within a void-dielectric system under square waveform applied voltage stress", *IET Science, Measurement & Technology*, vol. 8, no.2, pp. 81–88, 2014.

[KIM 17] Kim D.W., Oh H., Youn B.D., Kwon D., "Bivariate lifetime model for organic light-emitting diodes", *IEEE Transactions on Industrial Electronics*, vol. 64, no. 3, pp. 2325–2334, 2017.

[LAH 13] Lahoud N., Faucher J., Malec D., Maussion P., "Electrical aging of the insulation of low voltage machines: Model definition and test with the design of experiments", *IEEE Transactions on Industrial Electronics*, vol. 60, no. 9, pp. 4147–4155, 2013.

[MAL 04] Malerba D., Esposito F., Ceci M., Appice A., "Top-down induction of model trees with regression and splitting nodes", *IEEE Transactions on Pattern Analysis and Machine Intelligence*, vol. 26, no. 5, pp. 612–625, 2004.

[MAZ 09] Mazzanti G., "The combination of electro-thermal stress, load cycling and thermal transients and its effects on the life of high voltage ac cables", *IEEE Transactions on Dielectrics and Electrical Insulation*, vol. 16, no. 4, pp. 1168–1179, 2009.

[MON 11] Montanari G.C., "Bringing an insulation to failure: The role of space charge", *IEEE Transactions on Industrial Electronics*, vol. 18, no. 2, pp. 339–364, 2011.

[MON 89] Montanari G.C., Cacciari M., "A probabilistic life model for insulating materials showing electrical thresholds", *IEEE Transactions on Dielectrics and Electrical Insulation*, vol. 24, pp. 127–137, 1989.

[MON 02] Montanari G.C., Mazzanti G., Simoni L., "Progress in electrothermal life modeling of electrical insulation during the last decades", *IEEE Transactions on Dielectrics and Electrical Insulation*, vol. 9, no. 5, pp. 730–745, 2002.

[MYE 02] Myers R.H., Montgomery D.C., *Response Surface Methodology*, Wiley, 2002.

[PER 92] Persson E., "Transient effects in application of PWM inverters to induction motors", *IEEE Transactions on Industry Applications*, vol. 28, no. 5, 1992.

[PIC 13] Picot A., Malec D., Maussion P., "Improvements on lifespan modeling of the insulation of low voltage machines with response surface and analysis of variance", in *Proc. 9th IEEE International Symposium on Diagnostic for Electrical Machines, Power Electronics and Drives*, 2013.

[PRE 14] Preetha P., Thomas M.J., "Life estimation of electrothermally stressed epoxy nanocomposites", *IEEE Transactions on Dielectrics and Electrical Insulation*, vol. 21, no. 3, pp. 1154–1160, 2014.

[QI 11] Qi B., Sun Y., Hu W., Ding X., "A multi-stress accelerated life tests method for smart electricity meter based upon the life-stress model", in *Proc. 9th International Conference on Reliability Maintainability and Safety (ICRMS)*, pp. 1136–1140, 2011.

[SAL 15] Salameh F., Picot A., Chabert M., Maussion P., "Regression methods for improved lifespan modeling of low voltage machine insulation", *Mathematics and Computers in Simulation*, vol. 131, pp. 200–216, 2015.

[SAL 17] Salameh F., Picot A., Chabert M., Maussion P., "Parametric and non-parametric models for lifespan modeling of insulation systems in electrical machines", *IEEE Transactions on Industry Applications*, vol. 53, no. 3, pp. 3119–3128, 2017.

[SAL 19] Salameh F., Al Haddad A., Picot A., Canale, L., Zissis G., Chabert M., Maussion P., "Modeling the luminance degradation of OLEDs using design of experiments (DoE)", *IEEE Transactions on Industry Applications*, 2019.

[STO 93] Stone C.G., "The statistics of aging models and practical reality", *IEEE Transactions on Electrical Insulation*, vol. 28, no. 5, 1993.

[STO 05] Stone, G.C., "Partial discharge and electrical equipment insulation condition assessment", *IEEE Transactions on Dielectrics and Electrical Insulation*, vol. 12, pp. 891–904, 2005.

[SUT 11] Sut N., Simsek O., "Comparison of regression tree data mining methods for prediction of mortality in head injury", *Expert Systems with Applications*, vol. 38, no. 12, pp. 15534–15539, 2011.

[TAG 87] Taguchi G., Konishi S., *Orthogonal arrays and linear graph*, American Supplier Institute Press, Michigan, 1987.

[TAV 08] Tavner P.J., "Review of condition monitoring of rotating electrical machines", *IET Electric Power Applications*, vol. 2, no. 4, pp. 215–247, 2008.

[WEI 05] Weisberg S., *Applied linear regression*, Wiley, 2005.

[ZER 13] Zernig A., Bluder O., Spöck G., "Optimal design of experiments semiconductor lifetime data", in *9th Conference on Ph.D. Research Microelectronics and Electronics (PRIME)*, pp. 285–288, 2013.

[ZHA 14] Zhang J., Li W., Cheng G., Chen X., Wu H., Herman Shen M.H., "Life prediction of OLED for constant-stress accelerated degradation tests using luminance decaying model", *Journal of Luminescence*, vol. 154, pp. 491–495, 2014.

[ZHA 12] Zhang J., Liu F., Liu Y., Wu H., Wu W., Zhou A., "A study of accelerated life test of white OLED based on maximum likelihood estimation using lognormal distribution", *IEEE Transactions on Electron Devices*, vol. 59, no. 12, pp. 3401–3404, 2012.

[ZHU 14] Zhu X., Yan B., Chen L., Zhang R., Quan L., Mo L., "Multi-objective optimization design of a magnetic planetary geared permanent magnet brushless machine by combined design of experiments and response surface methods", *IEEE Transactions on Magnetics*, vol. 50, no. 11, seq. no. 8204004, 2014.

List of Authors

Yassine AMIRAT
ISEN YNCREA OUEST
Brest
France

Mohamed BENBOUZID
University of Brest
France

Marie CHABERT
INP-ENSEEIHT
Toulouse
France

Claude DELPHA
Paris Sud University
Gif-sur-Yvette
France

Demba DIALLO
Paris Sud University
Gif-sur-Yvette
France

Elhoussin ELBOUCHIKHI
ISEN YNCREA OUEST
Brest
France

Shahin HEDAYATI KIA
University of Picardy Jules Verne
Amiens
France

Mohammad HOSEINTABAR MARZEBALI
Shahrood University of Technology
Iran

Pascal MAUSSION
INP-ENSEEIHT
Toulouse
France

Antoine PICOT
INP-ENSEEIHT
Toulouse
France

Hubert RAZIK
Claude Bernard University
Lyon
France

Abdenour SOUALHI
LASPI Laboratory
Jean Monnet University
Roanne
France

Index

S, T, V

Summary of Volume 2

Chapter 2. Signal Processing Techniques for Transient Fault Diagnosis

José Alfonso Antonino DAVIU and Roque Alfredo Osornio RIOS

Chapter 3. Accurate Stator Fault Detection in an Induction Motor Using the Symmetrical Current Components

Monia BOUZID and Gérard CHAMPENOIS

Chapter 4. Bearing Fault Diagnosis in Rotating Machines

Claude DELPHA, Demba DIALLO, Jinane HARMOUCHE, Mohamed BENBOUZID, Yassine AMIRAT and Elhoussin ELBOUCHIKHI

Chapter 5. Diagnosis and Prognosis of Proton Exchange Membrane Fuel Cells

Zhongliang LI, Zhixue ZHENG and Fei GAO

Other titles from

ISTE

in

Energy

2020

BOISGIBAULT Louis, AL KABBANI Fahad
Energy Transition in Metropolises, Rural Areas and Deserts

2019

BENALLOU Abdelhanine
Energy Transfers by Convection
(Energy Engineering Set – Volume 3)
Energy Transfers by Radiation
(Energy Engineering Set – Volume 4)
Mass Transfers and Physical Data Estimation
(Energy Engineering Set – Volume 5)

LACHAL Bernard
Energy Transition

2018

BENALLOU Abdelhanine
Energy and Mass Transfers: Balance Sheet Approach and Basic Concepts (Energy Engineering Set – Volume 1)
Energy Transfers by Conduction (Energy Engineering Set – Volume 2)

JEMEÏ Samir
Hybridization, Diagnostic and Prognostic of Proton Exchange Membrane Fuel Cells: Durability and Reliability

RUFFINE Livio, BROSETA Daniel, DESMEDT Arnaud
Gas Hydrates 2: Geoscience Issues and Potential Industrial Applications

VIALLET Virginie, FLEUTOT Benoit
Inorganic Massive Batteries (Energy Storage – Batteries, Supercapacitors Set – Volume 4)

2017

BROSETA Daniel, RUFFINE Livio, DESMEDT Arnaud
Gas Hydrates 1: Fundamentals, Characterization and Modeling

LA SCALA Massimo
From Smart Grids to Smart Cities: New Challenges in Optimizing Energy Grids

MOLINA Géraldine, MUSY Marjorie, LEFRANC Margot
Building Professionals Facing the Energy Efficiency Challenge

SIMON Patrice, BROUSSE Thierry, FAVIER Frédéric
Supercapacitors Based on Carbon or Pseudocapacitive Materials (Energy Storage – Batteries, Supercapacitors Set – Volume 3)

2016

ALLARD Bruno
Power Systems-on-Chip: Practical Aspects of Design

ANDRÉ Michel, SAMARAS Zissis
Energy and Environment

DUFOUR Anthony
Thermochemical Conversion of Biomass for the Production of Energy and Chemicals

2015

CROGUENNEC Laurence, MONCONDUIT Laure, DEDRYVÈRE Rémi
Electrodes for Li-ion Batteries
(Energy Storage – Batteries, Supercapacitors Set – Volume 2)

LEPRINCE-WANG Yamin
Piezoelectric ZnO Nanostructure for Energy Harvesting
(Nanotechnologies for Energy Recovery Set – Volume 1)

ROBYNS Benoît, FRANÇOIS Bruno, DELILLE Gauthier, SAUDEMONT Christophe
Energy Storage in Electric Power Grids

ROSSI Carole
Al-based Energetic Nanomaterials
(Nanotechnologies for Energy Recovery Set – Volume 2)

TARASCON Jean-Marie, SIMON Patrice
Electrochemical Energy Storage
(Energy Storage – Batteries, Supercapacitors Set – Volume 1)

2013

LALOUI Lyesse, DI DONNA Alice
Energy Geostructures: Innovation in Underground Engineering

2012

BECKERS Benoit
Solar Energy at Urban Scale

ROBYNS Benoît, DAVIGNY Arnaud, FRANÇOIS Bruno, HENNETON Antoine, SPROOTEN Jonathan
Electricity Production from Renewable Energies

2011

GAO Fei, BLUNIER Benjamin, MIRAOUI Abdellatif
Proton Exchange Membrane Fuel Cell Modeling

MULTON Bernard
Marine Renewable Energy Handbook

2010

BRUNET Yves
Energy Storage

2009

SABONNADIÈRE Jean-Claude
Low Emission Power Generation Technologies and Energy Management

SABONNADIÈRE Jean-Claude
Renewable Energy Technologies

Printed and bound by CPI Group (UK) Ltd, Croydon, CR0 4YY